# AN INTRODUCTION TO THE
# ELECTROMAGNETIC WAVE

## F. A. WILSON

**BERNARD BABANI (publishing) LTD**
**THE GRAMPIANS**
**SHEPHERDS BUSH ROAD**
**LONDON W6 7NF**
**ENGLAND**

## Please Note

Although every care has been taken with the production of this book to ensure that any projects, designs, modifications and/or programs etc. contained herewith, operate in a correct and safe manner and also that any components specified are normally available in Great Britain, the Publishers and Author do not accept responsibility in any way for the failure, including fault in design, of any project, design, modification or program to work correctly or to cause damage to any other equipment that it may be connected to or used in conjunction with, or in respect of any other damage or injury that may be so caused, nor do the Publishers accept responsibility in any way for the failure to obtain specified components.

Notice is also given that if equipment that is still under warranty is modified in any way or used or connected with home-built equipment then that warranty may be void.

© 1993 BERNARD BABANI (publishing) LTD

First Published — March 1993

British Library Cataloguing in Publication Data
Wilson, F. A.
    Introduction to the Electromagnetic Wave
    I. Title
    537

ISBN 0 85934 315 4

Printed and Bound in Great Britain by Cox & Wyman Ltd, Reading.

# Preface

*Knowledge comes, but wisdom lingers.*

Alfred Lord Tennyson
(Locksley Hall)

Herein we discuss one of the most fascinating phenomena controlling our lives, perhaps the most important for the electromagnetic wave brings us both light and warmth from the sun. Without it we would be in darkness and even the stars at night would not be seen. Moreover the wave allows us to communicate over enormous distances and certainly without it society would collapse — in fact it never would have got started.

Without doubt the electromagnetic wave is a complicated affair and to fully understand it is beyond most of us. The light with which we see in all its colours and the man-made radio transmissions are all travelling at an unparalleled speed such that a wave could encircle the earth seven times in only one second. Even more frustrating is the fact that a host of these waves is buzzing around us incessantly. Yet with so much activity in the air around us we feel nothing of it. What is more each wave eschews its neighbours because they are of different frequencies. To us therefore the electromagnetic wave is one of Nature's specialities, designed for our use but unfortunately one which we will never see and have yet to fully understand.

Most text books treat the underlying principles of the electromagnetic wave as something requiring intense mathematical endeavour or alternatively as something to be dismissed in a paragraph or two. Here we try to fill the gap in between, never forgetting that unless we are careful, explanations can become formidable, therefore the aim is to make them as painless as possible. Remember that we are only scratching the surface of something very complex and perhaps that makes it all the more challenging and interesting.

Only a basic knowledge of electronics is assumed and the mathematics do not go beyond the ordinary equation.

Perhaps the more advanced readers will bear with us if it seems that we labour some points in an effort to help the less initiated.

*F. A. Wilson*

# Contents

# Chapter 1

# ELECTROMAGNETICS

World-wide communication on its own is enough to encourage us to understand the radio wave. But modern technology goes much further, we now have to learn about wave guides, optical fibres, radar, heating and many other wonders of electronics in which the electromagnetic wave plays a major part. Before we begin however it is essential to appreciate that the electromagnetic wave is a most complex affair altogether and we cannot study its fundamental structure unless we are first sure of the foundations on which it stands. On the basis that a little revision never comes amiss, this chapter discusses the basic theoretical considerations on which the remainder of the book rests. Skip over it if you dare!

## 1.1 Defining the Unseen

Perhaps the greatest difficulty we have in trying to get to grips with the electromagnetic wave arises from the fact that it has no effect whatsoever on any of our senses. How much easier it is to feel we know about a motor car or a clock for example since they can be seen. Yet there is already an unseen wonder at work in them, it is the *energy* which pushes the car along or turns the hands of the clock.

### 1.1.1 Energy

Energy is Nature's prime mover in life, a quantity rather difficult to pin down. For an explanation which will satisfy the physicist it is necessary to define both *force* and *work* first. Energy, force and work might be classed as invisible and intangible somethings yet we do manage to define them for example, as follows:

Force is most easily described as an influence and technically it is one which produces a change in the velocity of an object. Change in velocity implies acceleration, hence as an equation:

$$\text{force } F = m \times a$$

1

where $m$ is the mass of a body and $a$ is the acceleration the force produces. The SI unit of force is the *newton* (N) which is that force which applied to a mass of 1 kg gives it an acceleration of $1 \text{ m/s}^2$. One newton is equivalent to just over 100 grams (3½ oz) weight.

Work follows from force. When a force moves a body, work is done. It is measured by the product of the force and the distance its point of application moves, i.e.:

$$\text{work done } W = F \times s \text{ joules (J)}$$

where $s$ is the distance.

Energy follows from work. It is something capable of generating action, more precisely it is the ability of matter or *radiation* to do work. It is therefore explained more as what it does than what it is. Energy is present in all electronic activity. Since it is the capacity for doing work, it is measured in the same unit.

Energy comes in three forms, *potential*, *rest* and *kinetic*. The last is the energy of or due to motion and concerns us most here because much of the analysis we carry out on the fact moving electron rests on the energy it possesses. The symbol we use for kinetic energy is $E_K$ and the kinetic energy of a moving body depends on its mass and velocity only:

$$E_K = \frac{1}{2}m \times v^2 \text{ joules}$$

where $m$ is the mass in kg and $v$ the velocity in m/s.

In considerations involving minute particles such as electrons, the joule is an inconveniently large unit hence a unit referring to a single electron may be used, it is known as an *electron-volt* (eV). This is defined as the energy acquired by an electron when it is accelerated through a potential difference of one volt. The work done by the electric field (and hence energy gained by an electron) is given by $e \times V$ where $e$ is the electron charge ($1.602 \times 10^{-19}$ coulombs – see the next section) and $V$ is the accelerating potential difference (in this case 1 V). From this:

$$1 \text{ eV} = 1.602 \times 10^{-19} \text{ coulomb-volts, i.e. joules.}$$

Electron energies can therefore be quoted in this unit, usually resulting in convenient figures rather than in joules where almost inevitably 10 to some negative power is included. Note that the electron-volt is a unit of energy, not voltage.

### 1.1.2 *Charge*

Nothing in this world exists without *charge*. It is one of Nature's invisible entities possessed by atomic particles and is the driving force of electricity. There are two different kinds of charge which historically have been labelled *positive* and *negative*. Unlike charges give rise to an attractive force on each other, similar charges repel: the short golden rule is:

"like charges repel, unlike attract".

A charge is in fact a quantity of electrical energy and the electron charge is sometimes known as the *elementary charge* because so far nothing smaller has been confirmed. The electron charge ($e$) is $1.602 \times 10^{-19}$ coulombs.

Charles Augustin de Coulomb (the French engineer and physicist) first developed the relationship between charges *at rest* e.g. Figure 1.1(ii). His law states that the mutual force of repulsion between like charges or of attraction between unlike charges concentrated at points in an isotropic medium (the same throughout) is proportional to the product of the charges, and inversely proportional to the square of the distance between them and to the permittivity of the medium. As a formula:

$$F = \frac{Q_1 Q_2}{4\pi\epsilon d^2} \text{ newtons}$$

where $Q_1$ and $Q_2$ represent the charge magnitudes in coulombs, $d$ is the distance between their centres in metres and $\epsilon$ is the permittivity of the medium.

*Permittivity* is a measure of the ability of a material to store electrical energy when permeated by an electric field. It

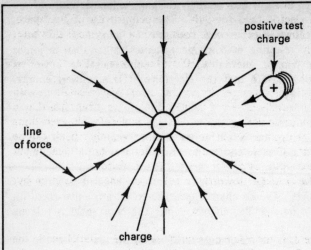

*(i)   field of a single (negative) charge*

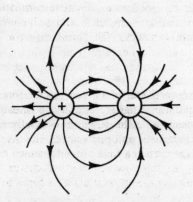

*(ii)   field between two charges of opposite sign*

**FIG. 1.1   LINES OF FORCE**

4

is given in a practical sense by the electric flux developed for a given electric field strength. The permittivity of free space, also known as the *electric constant* has the symbol $\epsilon_0$ where:

$$\epsilon_0 = 8.854 \times 10^{-12} \text{ farads per metre.}$$

$\epsilon_0$ is a constant necessary to link theoretical calculations with experimentally observed values. Free space means just that, a vacuum, free not only of atoms and molecules but everything else. This implies that no gravitational or other fields can be present. Completely free space is therefore not attainable but the concept is useful for theoretical calculations.

The *absolute permittivity* ($\epsilon$) of any material is given by:

$$\epsilon = \epsilon_r \epsilon_0$$

where $\epsilon_r$ is the *relative permittivity* of the material and in the case of a capacitor for example, is the measure by which the capacitance is increased when a particular material is substituted for air or free space. Relative permittivities range from less than 10 for mica, glass and polystyrene to over 1000 for certain ceramics. The term *dielectric constant* is also used.

### 1.1.3 Field

A field is simply a region of influence, therefore containing energy, consequently it has the potential to create some sort of force. In electronics the fields most usually encountered are the electric, magnetic and electromagnetic. As an example, an electric charge creates a field around it which has no effect whatsoever on anything except another charge. The field therefore has the capability of applying a force to the second charge.

To visualize a field and to demonstrate the existence of one on paper we use *lines of force* (electric or magnetic), drawing them closer together when the field is stronger. Figure 1.1 illustrates at (i) the field arising from a single charge and at (ii) the field arising from two charges of opposite polarity. Although strictly a field has no "direction", we give it one and this is indicated by arrows on the lines. For an electric field the direction is said to be that in which a free positive charge would move; for a magnetic field it is the direction in which a free North pole would move (but note that in practice a free North pole cannot exist).

For an electric field the *electric field strength E* is expressed in terms of the force $F_E$ it exerts on a unit charge. The force is therefore related to the electric field strength by:

$$E = F_E/e \text{ newtons per coulomb}$$

i.e. $F_E = eE$ newtons where $e$ is the electron charge in coulombs.

Magnetic field strength is discussed later in Section 1.2.

### 1.1.4 The Magnetic Force

Magnetism is not something on its own, it is the result of *moving* electric charges. It is important to be sure of this because when we study the electromagnetic wave later it will be realized that not only are the electric and magnetic fields linked in some way, they are in fact in such close partnership that one cannot exist without the other. A simple proof of this is by considering two current-carrying wires placed close together as shown in Figure 1.2. In (i) currents flow along the wires in the same direction and as the sketch indicates there is an attractive force between them. In (ii) the currents are in opposite directions and the wires repel each other (this form of current is known as *conduction current*).

Neither wire exhibits a charge when current flows in it so the attraction and repulsion cannot be that normally associated with electric charges. Because the wires are unaffected when no current flows in them, the forces which arise when currents flow must somehow originate from the charges in motion. We interpret these forces as the magnetic effect of current flow. Charge is one of Nature's *fundamental* resources but magnetism is not because it arises from charge.

To understand a little more of how charges in motion produce the so-called magnetic effect we must enlist the help of early scientists. Albert Einstein's revelation that a correction to Sir Isaac Newton's equations of motion was needed took the scientific world by surprise in the early nineteen hundreds.

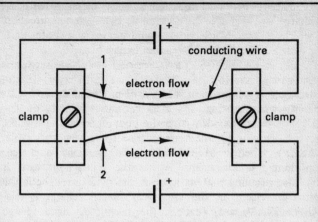

*(i)   currents in the same direction attract each other*

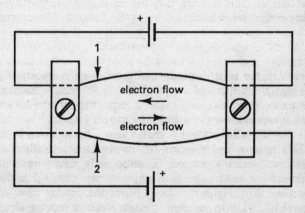

*(ii)   currents in opposite directions repel each other*

### FIG. 1.2   FORCES BETWEEN WIRES CARRYING CURRENTS

He decided that the mass of a body increases with its velocity:

$$m = \left( \frac{m_0}{\sqrt{1 - (v^2/c^2)}} \right)$$

where $m_0$ is the mass of a body when stationary and $v$ its velocity when moving. $c^2$ is an enormous velocity of $9 \times 10^{16}$ m/s so it is evident that the body must be moving like greased lightning for the mass increase to be even noticeable. Einstein's Theory of Relativity also contains one very important pronouncement — that the velocity of light is absolute, meaning that its value is the same throughout the entire universe. In the same way, electric charge is relativistically invariant, that is, the magnitude of a charge remains constant whatever its velocity.

Then along came the Dutch physicist H. A. Lorentz who suggested that material bodies and distances shorten in the direction of motion when they are moving, controlled by a rather similar relationship known as the *Lorentz Contraction:*

$$L = L_0 \sqrt{1 - (v^2/c^2)}$$

where $L$ is the length measured with the body in motion, $L_0$ is its length at rest and $v$ is the velocity of *relative* motion. The word "relative" is important, it means the net or resultant motion between the body and the observer. Again $c^2$ ensures that the effect is extremely small unless $v$ is very, very high.

This is only one example of the revolutionary scientific endeavour which went on in those early days but it is sufficient for us to appreciate why our two wires act in the way they do. Figure 1.3 looks inside the parallel wires of Figure 1.2. Let us see how a single electron moving along conductor 1 at a speed of $v$ [Fig.1.3(i)] from left to right is affected. Because the wire currents are in the same direction, electrons in conductor 2 are also moving from left to right at a speed of $v$, hence there is no relative motion. On the other hand the positive ions in conductor 2 are fixed, hence our electron in conductor 1 is passing each of them at a speed of $v$.

8

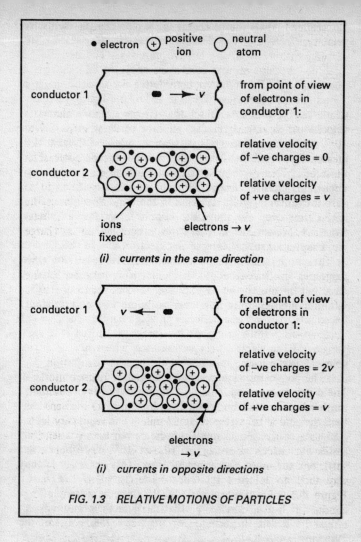

**FIG. 1.3    RELATIVE MOTIONS OF PARTICLES**

We can take this relative speed of *v* into account through the Lorentz contraction, accordingly from the point of view of the electrons in conductor 1 the distances between the positive charges in conductor 2 are less because of the contraction so a

9

greater concentration of charge arises (the charges themselves are unaffected). This leads to an *attractive electric force* arising entirely from the relative motions. The net effect is doubled because similar conditions apply to all electrons in conductor 2. Furthermore, considering the flow of charge in conductor 1 to be reversed as in (ii) of the figure, each electron in this conductor finds that (i) the negative charges in conductor 2 have a relative velocity of $2v$ whereas (ii) the positive charges have a relative velocity of $v$. It is clear that (i) leads to a repulsion double that of the electric force above whereas (ii) leads to an attraction equal to it, the net result being a *repulsive electric force* of similar magnitude to the attractive force for the currents in the same direction. These forces are over and above the normal interaction of charges and are in fact the *magnetic force* between the two wires.

Unquestionably the force per electron due to the Lorentz contraction is infinitesimally small however even for a few amperes of current up to $10^{20}$ electrons may be involved and when this figure is expressed as one hundred million million million, it is now easy to appreciate how relativity accounts for the force between moving charges which we call magnetism. Note also that the (magnetic) force is at right angles to the electric field within each wire which is driving the electrons along.

The above shows how current sets up a magnetic force or field and it is perhaps obvious that because (i) the total charge moving along the wire is equal to the sum of all the individual charges and (ii) that the current is simply an expression of the number of charges moving along in a given time (one ampere is equivalent to an average of $6.24 \times 10^{18}$ electrons passing per second), then the *magnetic flux density* ($B$) is directly proportional to the magnitude of the current ($I$), i.e. $B \propto I$.

In Figure 1.3 it is assumed that the space between the wires is air. If not air but some other medium then another term must be added to account for the *permeability* ($\mu$) of the medium, i.e. $B \propto \mu I$.

Permeability is a constant for most materials. It expresses the ease with which a magnetic flux is set up when a magnetic force (the *magnetic field strength*, $H$) is applied. As an equation:

$$\mu = B/H$$

The permeability of free space, also known as the *magnetic constant* has the symbol $\mu_0$ where:

$$\mu_0 = 4\pi \times 10^{-7} \text{ henrys per metre.}$$

As with permittivity (Sect.1.1.2) $\mu_0$ is a constant necessary to link theoretical calculations with experimentally observed values.

The *absolute permeability* ($\mu$) of any material is given by:

$$\mu = \mu_r \mu_0$$

where $\mu_r$ is the *relative permeability* of the material. Values of $\mu_r$ range from a little less than 1 up to many thousands. In many of the applications we consider here however, $\mu_r$ is aproximately equal to 1 hence $\mu$ is not appreciably different from $\mu_0$.

## 1.2 Electromagnetic Interaction

Through the theory of relativity Section 1.1.4 shows how an electric field, by moving charges along a wire produces what we call a magnetic field. This simple explanation will not illustrate all the ideas we examine but does give confidence in the fact that electric and magnetic forces cannot be separated, both arise from the *electromagnetic interaction* between charges. It remains to show that the interaction goes one step further in that whereas an electric field gives rise to a magnetic field, it is also a fact that a magnetic field can give rise to an electric field.

This can be demonstrated quite simply by recourse to Michael Faraday's early work. He found that moving a wire in a magnetic field induced an electromotive force (e.m.f.) between the ends of the wire. If an e.m.f. is so induced, there must be an electric field running the length of the wire affected by the magnetic field. It need not be the wire which is moved, it could equally be the field, i.e. it is the *relative* motion between a conductor and a magnetic field which is effective. The induced e.m.f., $E$ is given by:

$$\text{e.m.f. } E = \frac{\text{total flux cut}}{\text{time}} = \frac{d\phi}{dt}$$

where $d\phi$ is the change in the magnetic flux $\phi$ taking place during the very small period of time $dt$ – it is the rate of change of flux with time.

The following simple relationships prove the statement. Let a conductor of length $L$ move in a field of magnetic flux density $B$ and consider a single charge $Q$ (such as that of an electron) within the conductor. The charge moves with a velocity $v$ as shown in Figure 1.4. It is an experimental fact and one which can be proved theoretically that the magnetic force $F_M$ exerted on the charge is proportional to its velocity, i.e.:

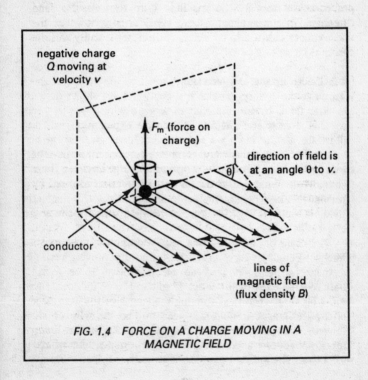

FIG. 1.4   FORCE ON A CHARGE MOVING IN A
MAGNETIC FIELD

$$F_M = Qv \sin \theta$$

where $v$ is at an angle $\theta$ with respect to $B$ as illustrated in the figure.

We have already defined the force $F_E$ on a charge in an electric field as $F_E = eE$ (Sect.1.1.3); by analogy that for the magnetic field follows as:

$$F_M = BQv \sin \theta$$

showing that the force on the charge $Q$ when it is moved through a magnetic field is proportional to the component of the velocity perpendicular to $B$. Thus when the velocity is perpendicular to $B$, i.e. $\theta = 90°$, then $\sin \theta = 1$ and $F = BQv$. The force is therefore maximum for motion perpendicular to $B$. Compared with the electric field therefore, for the magnetic field, $v$ has crept in so that the force is also controlled by the magnitude of $v$ and when the charge is stationary ($v = 0$), there is no force.

The work done on the charge equals force × distance (Sect.1.1.1). When the distance is equal to the length of the wire, $L$, then:

$$W = F \times L = BQLv \quad (v \text{ at right angles to } B).$$

Also the potential difference (p.d.) between two points is equal to the work done in moving a unit charge between these points, hence the p.d. across that part of the wire affected by the magnetic field = $W/Q$, i.e. p.d. = $BLv$.

In this consideration the p.d. across the ends of the wire is equal to the e.m.f. generated because no current flows. A p.d. across any two points implies that an electric field exists between them.

### 1.2.1 Enter Ampère and Faraday

So far all discussion has been in terms of electrons or other charges moving in a wire yet we know that radio waves do not need wires for their transmission. The question is therefore bound to arise at this stage as to how electromagnetic waves can exist in space or even brick walls. In true space for example, there are no charges around so how can the two types of field interact? Nature has provided for this and we first need to understand two laws:

### (i) *Faraday's Law of Electromagnetic Induction*
Consider a wire loop of area $A$ placed perpendicular to a magnetic field of magnetic flux density $B$ as in Figure 1.5.

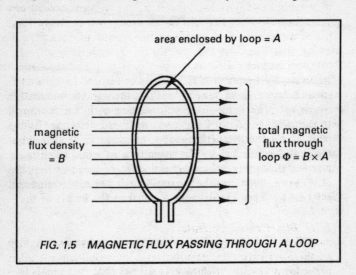

area enclosed by loop = $A$

magnetic flux density = $B$

total magnetic flux through loop $\Phi = B \times A$

FIG. 1.5   *MAGNETIC FLUX PASSING THROUGH A LOOP*

The magnetic flux density is the magnetic flux per unit area normal to the direction of the flux:

$$\text{magnetic flux density}, B = \frac{\text{magnetic flux } \phi}{\text{area } A}$$

from which $$\phi = B \times A$$

Michael Faraday discovered that the e.m.f. $E$ in such a loop is equal to the rate of change of flux through the loop, as already mentioned in Section 1.2, i.e.:

$$E = d\phi/dt.$$

14

The magnitudes of $B$ and $A$ do not affect the value of $E$, only the rate of change of either or both is important because $B$ and $A$ are part of $\phi$.

This of course is standard text book stuff. What is now of major importance is the fact that, as shown by James Clerk Maxwell's early work on electromagnetic theory, a metallic circuit is not essential for an e.m.f. to be generated in accordance with Faraday's law. The e.m.f. or *voltage gradient* can exist in any other material surrounding the changing flux, hence a magnetic field is able to give rise to an electric field.

## (ii) *Ampère's Law*

Again a closed path such as the loop in Figure 1.5 is considered and Andre Marie Ampère's law states that the summation of the magnetic field intensity around the path is proportional to the total current flowing across the surface bounded by the path. We generally think in terms of conduction currents but the law also refers to displacement and polarization currents. In these there is no flow of charge carriers, however they can give rise to the magnetic effects mentioned in the law just as if they were conduction currents.

### 1.2.2 Displacement Currents

A dielectric is an insulating medium or substance through which electricity can pass but not by conduction (unless breakdown occurs). When an electric field is applied to a dielectric the nuclei and the electrons are stretched out and each atom becomes a dipole as sketched in Figure 1.6(i). Polarization of the material follows as illustrated in (ii) with the atoms aligned parallel with the field. The electric field forces free charges to move to opposite sides of the dielectric and these set up an electric flux with the result that surface charges appear as shown. The small displacement of charges wihin each atom, all in the same direction, constitute a minute *displacement current* which only flows while polarization is being established.

If therefore a varying electric field permeates the dielectric, the displacement current, $I_{DIS}$ can be determined from the rate of change of the electric flux, $\Psi$ with respect to time, i.e.:

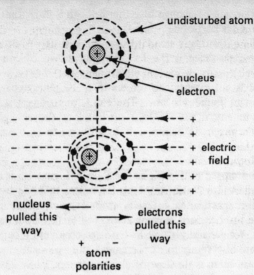

(i) *effect of an electric field on an atom*

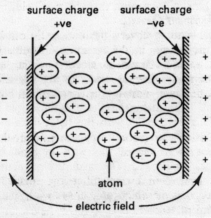

(ii) *polarization of a dielectric*

FIG. 1.6   DISPLACEMENT

$$I_{\text{DIS}} = d\Psi/dt$$

where $d\Psi$ is the change in electric flux occurring in the very short time $dt$. There is no displacement current if the electric flux remains constant.

From this it is evident that a magnetic field does not necessarily arise from the effects of a conduction current, it may in fact be the result of a displacement current.

### 1.2.3 Putting It All Together

From this chapter so far we realize that electromagnetism is about charges which move uniformly (e.g. steady current) for then magnetic forces and fields are involved. It is evident that whereas an electric field can give rise to a magnetic field, the converse is also true, i.e. a magnetic field can produce an electric field, moreover these effects are not confined to conduction wires but also occur in solid, liquid or gaseous substances. No stretch of the imagination is required to show that if an electric field produces a magnetic field which itself produces an electric field, then the two fields are linked together and if there are no losses, can be self sustaining. Also we have the general impression that the two fields must be at right angles to each other.

What we have not discovered is how all this can happen in space where no charges in the dielectric are present for the fields to act upon. Maxwell deduced from the laws above that electromagnetic waves can exist in free space also. His equations are too complicated for us to unravel so we must accept that:

"electromagnetic interaction occurs in any circuit path, even in a dielectric or space".

Ampère's Law is in terms of current flow. However current, whatever its nature, flowing in an impedance produces a voltage gradient, therefore an electric field. In his work Maxwell showed that even free space has an impedance and this is made up of the electric constant, $\epsilon_0$ (Sect.1.1.2) with the magnetic constant, $\mu_0$ (Sect.1.1.4). The *intrinsic impedance of free space*, $Z_0$ is considered in Appendix 2 and

17

has the value of 377 ohms.

A current flowing through $Z_0$ therefore produces a voltage according to Ohm's Law.

# Chapter 2

# THE ACCELERATING CHARGE

It is a fact which is both observed and also mathematically proven that electric and magnetic forces generally fall off according to the square of the distance between them. That is, when two opposing or attracting quantities are close together the effect is appreciable but when far apart the effect is negligible. An example is given by Coulomb's Law which shows that the mutual force between two point charges varies inversely as the square of the distance between them. If therefore the electromagnetic wave functioned in the same way, it would be pretty useless except at high transmitter powers and short distances. Fortunately the wave follows a different rule in that it falls off as the first power of the distance. Thus for most forces, increasing the distance between them a hundredfold reduces the effect to one ten thousandth but for the electromagnetic wave it is reduced to one hundredth. Hence as we well know, electromagnetic waves can cover large distances and still be useful. This, of course is a theoretical consideration (which we develop below), many earthly factors and tricks of the trade change things completely.

## 2.1 The Wave Is Born

When explaining how an electromagnetic wave comes about, the experts fall back on their *field equations*, originally developed by Maxwell. Not only are these equations somewhat complicated, they do not lead easily to an appreciation of the wave in a physical sense, so they are of little help in our seeing the wave in the mind's eye thereby developing an affinity for it. We need therefore to conjure up for ourselves some ideas which indicate the main underlying characteristics of the wave. Firstly we meet the problem of calculating the electric field developed by a *moving* charge. The difficulty can perhaps best be illustrated by looking up at the stars at night. Suppose there is one said to be 100 light years away (i.e. 100 times the distance light travels in one year). The

19

light emanating from the star has taken 100 years to reach our eyes and in that time the star can have moved to anywhere else in the universe or even have disappeared altogether. So we do not know where the star is even though we are looking at it!

In the same way if a charge $Q$ as shown in Figure 2.1(ii) is moving, we at P cannot determine the magnitude ($E$) of the

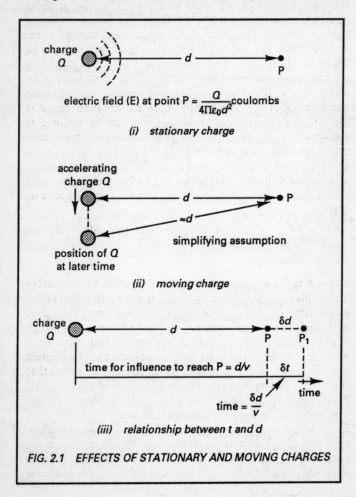

charge $Q$

electric field (E) at point P = $\dfrac{Q}{4\Pi\varepsilon_0 d^2}$ coulombs

*(i) stationary charge*

accelerating charge $Q$

position of $Q$ at later time

$\approx d$

simplifying assumption

*(ii) moving charge*

charge $Q$

$\delta d$

time for influence to reach P = $d/v$

$\delta t$

time = $\dfrac{\delta d}{v}$

time

*(iii) relationship between t and d*

FIG. 2.1 EFFECTS OF STATIONARY AND MOVING CHARGES

electric field it produces *at this very moment* but only at some time previously because of the delay arising from the time it takes for the influence of $Q$ to arrive at P. There is of course no problem with a stationary charge because once its influence arrives at P it is unchanging as shown in (i) of the figure and here Coulomb's Law applies. However when $Q$ is on the move its influence at P at any instant is delayed by a time $d/v$ where $v$ is the speed of travel. (We may need to recall from our school days that distance divided by velocity $(d/v)$ = time.) The delay is known as the *retarded time*.

Now when the full mathematical formula is developed in which the delay time is accounted for, it becomes difficult to see the wood for the trees. However the most likely condition is that the charge movement is over a small distance compared with $d$. Accordingly since the distance moved is small during the retarded time, this time can be assumed as constant for all charge positions as illustrated by Figure 2.1(ii) where the distance from P to the position of the charge at a later time is also reckoned as $d$. This assumption leads to a simplified formula from which we derive the electric field arising from an accelerating charge at time $t$ and distance $d$. It is:

$$\text{electric field}, E \text{ at P (time } t) = \frac{Q}{4\pi\epsilon_0 v^2} \times \frac{a(t - d/v)}{d}$$

where $a$ is the component of the acceleration of the charge on the plane perpendicular to $d$ and is the acceleration at the earlier time $(t - d/v)$. Not an easy formula to get to grips with because of its involvement with time but we must accept it as it stands without looking for the original work from which it is derived (this is far more complicated). The formula tells us firstly that $E$ is inversely proportional to $d$ now that the charge is *accelerating* whereas as shown in Figure 2.1(i) $E$ is inversely proportional to $d^2$ for the stationary charge. The term $a(t - d/v)$ is known as the *retarded acceleration* i.e. it is the acceleration at the earlier time when the influence left $Q$. This can be summed up by saying that if we watch the charge from a position P then the electric field we experience is proportional to the acceleration of the charge at

21

the earlier time $(t - d/v)$.

For a given $Q$ the first part of the equation is of fixed value and from the second part we can obtain a relationship between $t$ and $d$. Looking more closely at $a(t - d/v)$ it is evident that if $t$ is increased slightly to say, $t + \delta t$ (where $\delta t$ is a little time) then the effect of the field will be experienced a little further $(\delta d)$ from P at $P_1$ [see (iii) of the figure]. Put in other words, slightly later (i.e. adding the little time $\delta t$) for the formula to hold good a small distance $\delta d$ $(= v \delta t)$ must be added. The formula therefore shows that $t$ and $d$ are interchangeable, $t$ must increase because that is the way of things, hence $d$ also increases. Thus the electric field moves outwards from the source as time progresses at the velocity of $v$. Nature earlier decided on the value of $v$ and more recently Maxwell discovered how to calculate it, i.e. from:

$$v = \sqrt{(1/\mu_0 \epsilon_0)} = \sqrt{\frac{1}{(4\pi \times 10^{-7})(8.854 \times 10^{-12})}}$$

$$= 2.998 \times 10^8 \text{ metres per second}$$

(see Sects.1.1.2 and 1.1.4).

For many calculations it is more convenient to work to $3 \times 10^8$ m/s and it is found that for air the values of $\mu_0$ and $\epsilon_0$ are so close to 1 that the same figure can be used. This particular velocity is generally designated by the quantity symbol $c$.

From the above therefore we prove the salient characteristics of the electromagnetic wave to be:

(i)    any charge accelerating produces an electromagnetic wave;

(ii)   the strength of the wave varies inversely as the distance from its origin;

(iii)  the wave travels at a velocity $c$ ($3 \times 10^8$ m/s in a vacuum but less in other media). From this it is evident that electromagnetic waves require no material medium for their journeys.

## 2.2 Radiation

Having now decided that it is an *accelerating* charge which produces an electric field moving outwards at the speed $c$, we might usefully consider the acceleration conditions when a charge is oscillating. Clearly there is an acceleration in one direction followed by an acceleration in the opposite direction. In the formula for the electric field in Section 2.1 therefore $a$ may be positive or negative, the field polarities will therefore change accordingly. For simplicity of explanation consider how the electric field varies as it moves away from a sine wave oscillating charge. To answer this we must determine the acceleration pattern of the charge first and fortunately for a sine wave, the acceleration is also a sine wave as developed in Figure 2.2(i). Curve 1 represents the position of the charge at any instant. Its velocity is the rate of change of position which is given by the slope of Curve 1, resulting in Curve 2. The acceleration is the rate of change of velocity given by the slope of Curve 2 and shown as Curve 3, (the mathematicians among us will simply say that for the acceleration of the charge, the second differential is required which for a curve of sin x is $-$ sin x). From this we conclude that a charge oscillating according to a sine function generates a wave with an electric field strength varying with distance as shown in Figure 2.2(ii). The dotted curve indicates how the wave has progressed at a later time.

### 2.2.1 The Dipole Radiator

So far we have considered the effects of a single accelerated charge but not defined this charge which if it were a single electron at $1.602 \times 10^{-19}$ coulombs would give rise to an electric field somewhat feeble. However when an electric current flows millions of electrons are accelerated and the principle of superposition of fields shows that the total field is simply the sum of all the individual fields, hence in practice the total charge being accelerated could easily be many coulombs.

To demonstrate how the charges accelerate first in one direction then in the opposite, a dipole radiator may be used as shown in Figure 2.3. An oscillator is connected to a single wire split at the centre. If at any instant the polarity of the

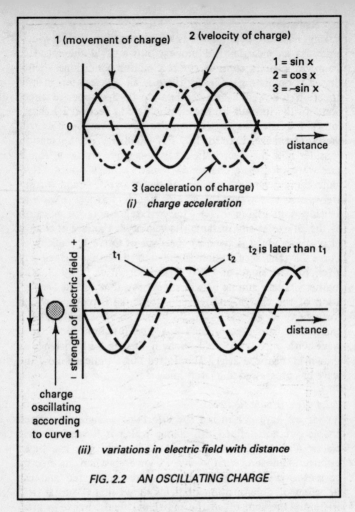

FIG. 2.2 AN OSCILLATING CHARGE

oscillator is as shown, then electrons are attracted out of the top wire and are also forced into the bottom wire, this is equivalent to an electron flow down the complete wire, i.e. in this particular case electrons are accelerated downwards.

On the next half-cycle of the oscillator the opposite condition holds so the electrons are accelerated upwards. This is

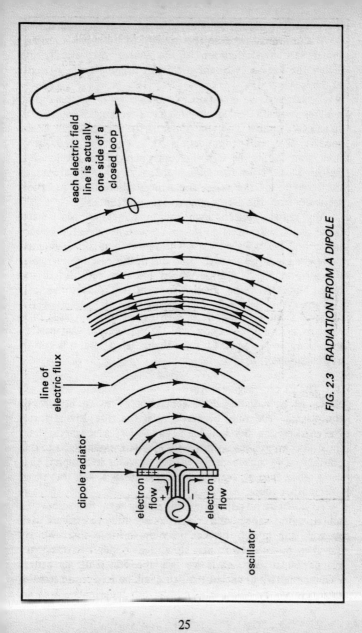

each electric field line is actually one side of a closed loop

line of electric flux

dipole radiator

electron flow

electron flow

oscillator

FIG. 2.3   RADIATION FROM A DIPOLE

the condition needed for the radiation of an electromagnetic wave and clearly because the strength of the wave is proportional to the acceleration of the charge, theoretically the higher the frequency of the oscillator, the greater the strength of the wave.

Considering one direction of propagation only (this dipole radiates equally in all horizontal directions), Figure 2.3 attempts to show how the lines of the electric field might appear. We recall from Section 2.1 that for any given direction from the dipole it is the component of the acceleration *on the plane perpendicular to that direction* which determines the strength of the wave, accordingly there is no radiation vertically from the dipole shown, only horizontally.

Any radiator has its own built-in inductance and capacitance which when excited by an alternating current, produces the normal fields we associate with tuned circuits. An antenna therefore gives rise to such fields but in this case the energy of the surrounding field is developed and then returned to the tuned circuit twice per cycle. The fact that it is returned and also that the field falls off inversely as the square of the distance from the antenna means that effectively most of it remains near to the source and takes practically no part in radiation. To distinguish it from the radiated field, it is known as the *induction field*.

## 2.3 The Electromagnetic Wave

So far we have discussed the generation of a wave in terms of the electric field only because this is the true driving force. Nevertheless as Section 1.2.3 concludes, a magnetic field must also be present with the two fields at right angles to one another. The coalescence of the two waves is so complete that each relies entirely on the other and so in a way they keep one another going.

To illustrate on paper something never seen, that something all pervading and varying possibly many millions of times each second, is daunting. Be that as it may, in Figure 2.4(i) we try. To show how the two fields are at right angles, two imaginary planes are drawn and if we face the oncoming wave it is evident that (ii) or (iii) of the figure will be seen, again remembering that the lines are purely our imaginative way of

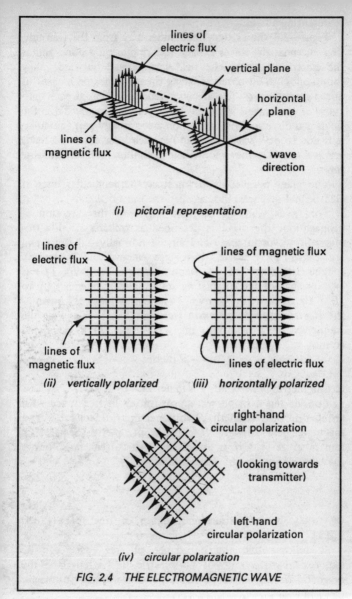

*(i) pictorial representation*

*(ii) vertically polarized*

*(iii) horizontally polarized*

*(iv) circular polarization*

FIG. 2.4   THE ELECTROMAGNETIC WAVE

representing a field.

Figure 2.3 shows that once under way from the transmitting antenna, the waves expand like ripples on a pond and as the electric and magnetic field lines spread outwards, they soon appear to an observer facing them as lines on a sheet of linear graph paper. The arrows reverse direction at each half-cycle of the wave. When, as in the case of Figure 2.4(i) the fields and direction of propagation are mutually perpendicular, it is said to be a *plane wave* and in such a wave the two fields are *in phase* (i.e. they reach their maximum levels at the same time).

The plane parallel to the mutually perpendicular lines of electric and magnetic flux is called the *wave front*.

For fields vibrating at right angles to the direction of propagation, the wave is described as *transverse*. Electromagnetic waves in space and air are normally transverse and are called TEM waves (transverse electric and magnetic).

The frequency of an electromagnetic radiation ($f$) is exactly that of the originating source and frequencies up to $10^{22}$ Hz or more are known. The entire range is known as the *electromagnetic spectrum* (see Sect.2.3.3). Knowing the frequency of a radiation, the wavelength ($\lambda$) is calculated from:

$$\lambda = c/f \text{ metres}$$

($c$ is in metres per second, $f$ is in hertz).

Considering a plane wave propagating in free space with an electric field strength, $E$ volts per metre (Sect.1.1.3) and magnetic field strength $H$ amperes per metre (Sect.1.1.4), remembering that $H$ is associated with current flow, brings the Ohm's Law type of relationship:

$$H = E/Z_0$$

where $Z_0$ is the intrinsic impedance of free space (Sect. 1.2.3).

It follows that the impedance of free space can be expressed by the ratio of the electric field strength to the magnetic field strength of an electromagnetic wave propagating through it.

The power density, $p$ in an electromagnetic wave in free space can therefore be calculated from the field strengths as:

$$p = E^2/Z_0 = H^2 Z_0 \text{ watts per square metre.}$$

### 2.3.1 Polarization

Knowing that an electric field can accelerate charges within a wire (such as a receiving antenna), it is possible to see from Figure 2.3 that for maximum effect (i.e. for maximum receiving antenna signal voltage), in this particular case the wire must be vertical because the transmitting dipole is vertical. By turning the dipole to horizontal the electric flux then cuts the wire at right angles and no longer drives electrons *along* the wire to produce an antenna current. Accordingly the receiving antenna must follow suit. Thus it is useful to be able to quote the field direction of any electromagnetic wave, this is known as describing its *polarization*.

To avoid ambiguity, it has been generally agreed that the polarization of a wave should be specified by its *electric* field. An advantage of this choice is that in the simple dipole system we have been studying and many others, since it is the electric field which does most of the work, then the receiving antenna will have the same orientation as that of the transmitter. In Figure 2.4(ii) therefore the wave is said to be vertically polarized because the lines of electric flux are vertical and it is horizontally polarized in (iii). A wave may not be truly vertical or horizontal, it could be radiated at some other angle or changes may be forced on it during its travels.

The *plane of polarization* is the plane which contains both the direction of the electric field and the direction of propagation.

Another polarization employed (but not with dipole antennas) is known as *circular* which is actually one particular type of *elliptical* polarization. In this the pair of fields continually rotates at the rate of one revolution per cycle. To indicate which way round the pair is going, on looking towards an oncoming wave, a clockwise rotation is described as *positive* or *right-handed* and anticlockwise as *negative* or left-handed as illustrated in Figure 2.4(iv).

## 2.3.2 The Receiving Dipole Antenna

This section contains a few abbreviated notes on a single type of receiving antenna to help us on our way. More on antennas follows in Chapter 5 for there are so many antenna shapes, sizes and types that a fuller discussion is essential. The wavelength of transmission is a dominant factor and generally the dipole pictured in Figure 2.3 is known as a *half-wave dipole* for its overall length is approximately equal to half the wavelength of the electromagnetic wave being radiated. It is slightly shorter than might at first be expected because the wavelength in the antenna rods is less than that in air owing to the lower velocity of the wave in media having a permittivity and/or permeability greater than 1 (Sect.2.1). The receiving dipole is constructed similarly.

An electromagnetic wave enveloping a receiving antenna might be considered as having two separate effects, that due to the electric field and that due to the magnetic field. When for example, a vertically polarized wave meets a vertical antenna, the latter is parallel to the electric flux and at right angles to the magnetic flux, both conditions resulting in maximum force on the antenna electrons. We recall that the electric field exerts a direct force on the electrons similarly the magnetic field in sweeping across the antenna wire also produces a force on them and in the same direction. However we cannot consider the two effects separately for the fields are simply two different offshoots of a single phenomenon. The effect of the magnetic field is therefore equally important as that of the electric field, nevertheless we normally find it more convenient to consider the action in terms of the electric component of the wave as follows.

If an electric potential is applied across the ends of a length of conducting wire then an electric field traverses the wire. As Section 1.1.3 points out, there is a force exerted on every electron within the wire ($F_E = eE$) and those free electrons capable of moving do so – this is the normal current as evaluated by Ohm's Law. Things are not so very different when an incoming electromagnetic wave permeates the wire for if the total electric field strength generated in the wire by the complete wave is $E$ volts per metre, then one metre of wire with the same polarization angle as the wave would

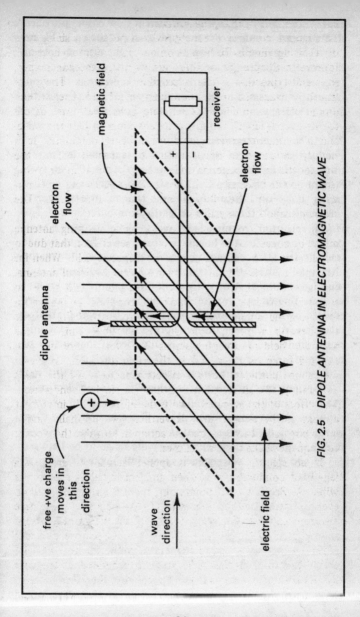

FIG. 2.5  DIPOLE ANTENNA IN ELECTROMAGNETIC WAVE

experience $E$ volts across it hence current flow ensues provided that a closed circuit across the wire exists as shown in Figure 2.5. This drawing is for one direction of the electric field of the wave only, as the cycle continues the wave of course reverses its direction. The strength of a wave is therefore frequently expressed in the unit volts per metre or frequently more practically in microvolts per metre ($\mu$V/m).

### 2.3.3 The Electromagnetic Spectrum

The earthly range of electromagnetic radiations extending from those at the longest wavelengths of some 100 km ($f = 3$ kHz) to the shortest, say $3 \times 10^{-5}$ nanometres ($f = 10^{22}$ Hz) is shown on Figure 2.6. Cosmic rays are of even shorter wavelengths and these arrive on earth from outer space. The *visible* spectrum, running from red to violet is only a small part of the whole, it is bounded on the lower frequency side (red) by the infra-red and on the higher frequency side by the ultraviolet. Above the ultraviolet are X-rays and gamma rays, the medical use of X-rays is well known, gamma rays are not so well known, they arise from nuclear reactions and are emitted by certain radioactive substances. Below the infra-red range are the wavelengths generally used in radio transmissions, subdivided as shown and as amplified at the bottom of the figure.

Frequently the term *microwave* arises, micro in this case not meaning one millionth as is generally found in electronics but here simply used to indicate *small*, i.e. microwaves are those of small (short) wavelength. The exact range of these waves seems to defy universal acceptance for many different ones are quoted. Here we consider the range to be between 1.3 m and 3 mm, i.e. frequencies from just below 1 GHz up to 100 GHz.

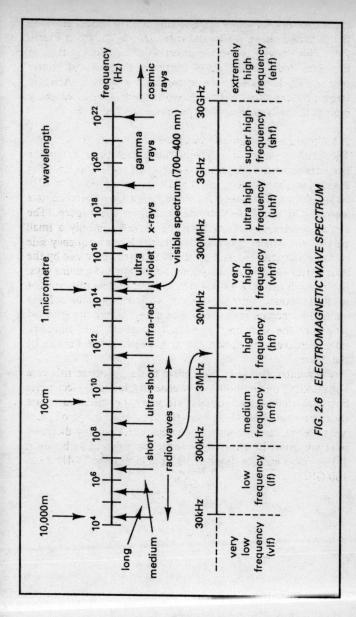

FIG. 2.6 ELECTROMAGNETIC WAVE SPECTRUM

33

# Chapter 3

# WAVE GEOMETRY

Just as we have contrived to illustrate a field by a collection of lines of force, so again on paper lines must be employed to indicate the direction in which an electromagnetic wavefront travels. The wave is therefore considered as a narrow ray. Such a notion while in no way capable of indicating the spread of the wave-front (e.g. km² for radio transmissions but less than a hair's breadth in an optical fibre) adequately illustrates how the waves are affected when they meet a boundary or travel in different substances. The true value of this chapter will be realized when we meet for example the ionosphere, waveguides and optical fibres.

Angles at which rays strike a surface are usually expressed relative to the *normal* rather than to the surface itself for this may not always be regular. The normal is simply a straight line drawn perpendicular to the surface at the point at which the ray strikes.

## 3.1 Reflection

Figure 3.1(i) indicates the general rules of reflection, commonly experienced in the practical world with light but also appropriate to other electromagnetic waves. The laws governing reflection where the incident and reflected waves are in the same medium are:

(i)  the incident, normal and reflected rays lie in the same plane, the two rays being on opposite sides of the normal;

(ii)  the angle of reflection ($\theta_r$) is equal to the angle of incidence ($\theta_i$), both angles being relative to the normal.

## 3.2 Refraction

When travelling in a medium other than a vacuum, electromagnetic waves do so at a reduced velocity:

$$v = c/n_c$$

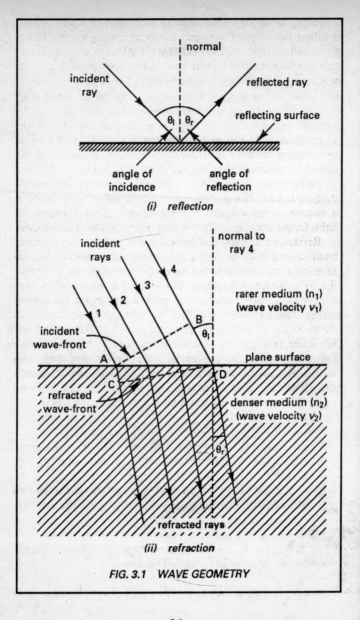

FIG. 3.1   WAVE GEOMETRY

where $n_c$ is the *complex index of refraction*, calculated from a rather heavy-going formula, involving amongst other things the conductivity of the medium. For most work however this can be considered to be zero and happily $n_c$ then reduces to $\sqrt{(\mu_r \epsilon_r)}$. Again very frequently $\mu_r = 1$ hence:

$$v = c/n$$

where $n$ is now the *index of refraction* or *refractive index* (no longer complex and is equal to $\sqrt{\epsilon_r}$ where $\epsilon_r$ is the dielectric constant, equal to $\epsilon/\epsilon_0$ (Sect.1.1.2).

Furthermore since $n = c/v$, the index of refraction can be defined for a medium as the ratio of the velocity of a wave in a vacuum to its velocity in the medium. Two examples for light frequencies are 1.33 for water and about 1.5 for glass.

Refraction of light is seen in everyday life as the apparent bending of a stick or ruler partly immersed in water. We can imagine how this happens from the simple idea in Figure 3.1(ii) in which AB represents a wavefront consisting of rays 1 − 4. Consider that ray 1 is about to enter the denser medium then clearly rays 2 − 4 have progressively further to travel in the rarer medium before they too strike the surface. While for example, ray 4 travels from B to D, ray 1 is advancing more slowly from A to C in the denser medium. This results in a *refracted wavefront* CD and as the sketch shows, the whole wave is bent towards the normal.

Simple trigonometry shows that:

$$\frac{\sin \theta_i}{\sin \theta_r} = \frac{BD}{AC}$$

and since BD and AC are proportional to the wave velocities $v_1$ and $v_2$ in the two mediums:

$$\frac{\sin \theta_i}{\sin \theta_r} = \frac{v_1}{v_2} = \frac{n_2}{n_1}$$

where $n_1$ and $n_2$ are the refractive indices of the incident and onward transmission regions respectively. This is known as Snell's Law (after the Dutch astronomer Willebrord Snell).

We need not be too concerned with practical values for generally our interest lies mainly with the fact that the velocity of an electromagnetic wave changes on entering a medium of differing refractive index and therefore the wave is refracted towards the normal when $n_1/n_2$ is less than 1 and away from the normal when it is greater than 1. There is no refraction when $n_1 = n_2$.

## 3.3 Total Internal Reflection

This is the curious condition of reflection which is attributable to refraction. Transmission of electromagnetic waves confined in glass fibres is only possible through this phenomenon. It only occurs when a wave passes from one medium to another with a lower index of refraction and the angle of refraction is equal to or exceeds 90°. Let us examine this in a practical way.

Consider Figure 3.2 in which the incident rays are within a block of glass of index of refraction, $n_1 = 1.5$. The glass is surrounded by air, index of refraction, $n_2 = 1$. Ray A arrives at the glass surface at an angle $\theta_i$ of 30° to the normal and is refracted into the air at an angle $\theta_r$ of 48.6° as calculated from Snell's Law ($n_1 \sin \theta_i = n_2 \sin \theta_r$). If now $\theta_i$ is increased, $\theta_r$ increases more and eventually as ray B shows, for $\theta_i = 41.81°$, $\theta_r = 90°$ and the refracted ray then travels along the interface between the two media. The value of $\theta_i$ which gives rise to this condition is known as the *critical angle of refraction* and values of $\theta_i$ above this result in *total internal reflection*. Rays such as C therefore cannot escape into the other medium (air in this case) and so are reflected within the glass according to the normal rules for reflection (Sect.3.2).

## 3.4 Diffraction

Contrary to what we know generally about electromagnetic waves, they *are* able to bend around the edge of an opaque object in their path — this is known as *diffraction*. It occurs when edges or corners of boundaries are traversed by a wave of much shorter wavelength than the size of the boundary.

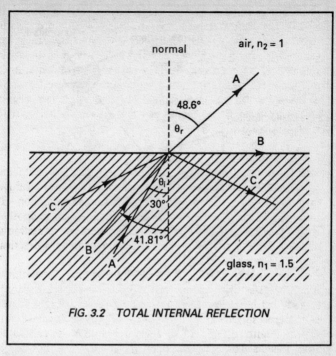

*FIG. 3.2   TOTAL INTERNAL REFLECTION*

The mathematical solutions of diffraction are very complex and so restrict us to the simple case of diffraction of a plane wave at the edge of a screen, nevertheless sufficient for an appreciation of what the result is.  This is that unexpectedly waves extend into what one would expect to be a clear-cut shadow region.  The effect is illustrated in Figure 3.3 which shows how the shadow on the screen of an opaque object has a fuzzy edge rather than the sharp edge which might be expected.  This is because waves spread out after leaving the object.  Christiaan Huygens (the Dutch scientist) in the late 1600's first put forward the idea that light travels in the form of waves and we are indebted to him for his principle.  This in brief states that every point on a wavefront can be considered as a point source of secondary wavelets which spread out in all directions, the wavefront at any time being the envelope of these wavelets.  This is effectively saying that the straight lines

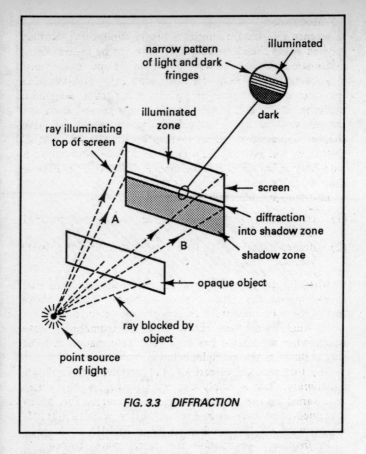

narrow pattern
of light and dark
fringes

illuminated

dark

illuminated
zone

ray illuminating
top of screen

A

B

screen

diffraction
into shadow zone

shadow zone

opaque object

ray blocked by
object

point source
of light

FIG. 3.3   DIFFRACTION

we use for illustrating light and other electromagnetic waves
should really have frayed or fluffy edges, not of course that
this is practicable.

The secondary wavelets of for example rays A and B on
the figure therefore are part of the diffraction into the shadow
zone and the light which has spread in consists of a narrow
pattern of light and dark fringes. All electromagnetic waves
give rise to diffraction which we may sum up as a slight
bending at any discontinuity in the medium through which
the wave is travelling.

40

## 3.5 Modulation

In essence this section considers briefly some of the various ways in which the electromagnetic wave can be used to carry information whether the wave be free in the open or confined in a waveguide or fibre. Here the term *information* implies the electrical waveforms of speech, music, data, television, facsimile, etc. Information involves change and there are several ways in which an electromagnetic wave can have changes impressed on it so that information is transmitted. Overall we can say that for a signal to contain information it must vary in an unpredictable manner (if it were predictable there would be no point in having it).

Information signals fall generally into two types:

(1) continuous signals which have any value between fixed limits; and

(2) discrete signals which have a known number of possible values.

Whatever the type of signal, it can usually be taken that at any instant the signal level on a channel must be above that of the noise present for the signal to be detected.

In what follows the *carrier* is an electromagnetic wave which when modulated has undergone deformation in some way as shown in the examples below.

The first example of category (1) to arrive was *amplitude modulation*. This is simply a carrier wave ($f_c$) with its amplitude varied by the information it has to carry. This is best explained graphically as in Figure 3.4(i) in which a 100 kHz carrier wave ($V_c$) is modulated by a 20 kHz wave ($V_m$).

In *frequency modulation* the carrier wave frequency is varied according to the amplitude of the modulating wave and at its frequency. The resulting wave therefore has constant amplitude but varying frequency as shown in Figure 3.4(ii) for a 250 kHz carrier modulated by a 20 kHz wave. Frequency modulation requires a greater bandwidth than amplitude modulation but has the advantage that it is less sensitive to interference because amplitude changes imposed on it have no effect.

With *phase modulation* the phase angle of the carrier wave is varied about its unmodulated value in accordance with the

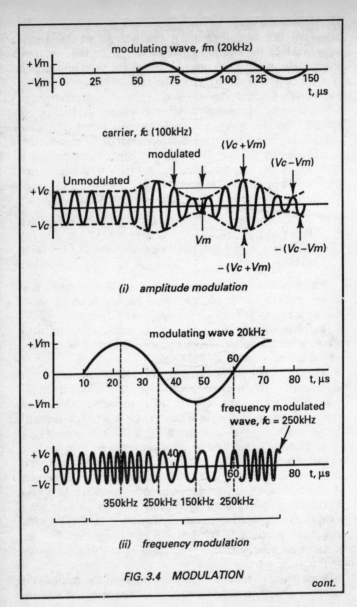

*(i) amplitude modulation*

*(ii) frequency modulation*

FIG. 3.4   MODULATION

cont.

42

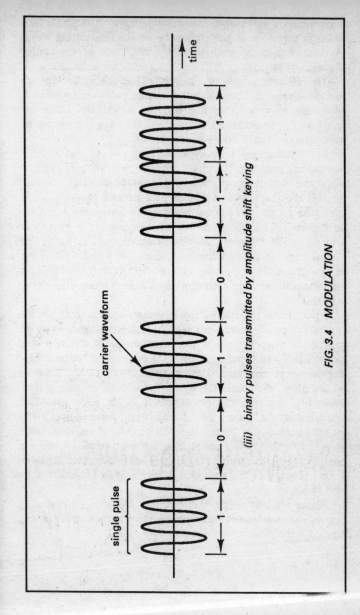

(iii) *binary pulses transmitted by amplitude shift keying*

FIG. 3.4  MODULATION

instantaneous value of the modulating wave. The phase variation is therefore at the frequency of the modulating wave. As with frequency modulation the amplitude of the carrier wave is not varied.

The discrete signals of category (2) as the name suggests are usually in the form of pulses. Most pulse transmission is in binary in which pulses of constant height and width are employed to indicate a binary 1 and no pulse (i.e. return to zero level) indicates a 0.

For transmitting a binary pulse train, say by radio, there are several methods available. The simplest yet sufficient to show the basic technique is known as *amplitude shift keying*. In this the amplitude of the electromagnetic carrier wave is switched on and off so generating a waveform such as shown in Figure 3.4(iii). In *frequency shift keying* the frequency is shifted rapidly between two fixed frequencies, i.e. one of the frequencies represents a binary 1, the other a binary 0. There is also the more complicated *phase shift keying* in which the two states differ from each other by a known phase difference, e.g. 180°.

Apart from the use of similar pulses to represent binary codes, the pulses themselves may be varied in height, width or position. For example in *pulse amplitude modulation* the amplitude of each pulse represents the voltage of the modulating wave at the instant of sampling. In *pulse duration modulation* the time durations or widths of the pulses represent the modulation, similarly with *pulse position modulation* the pulses are shifted from their normal time positions according to the modulation. *Pulse code modulation* is a further development in which the modulating wave is sampled regularly (tested for voltage level), each sampling level being allocated a binary code which is then transmitted as shown above.

Above are only a few of the various methods for modulating an electromagnetic wave carrier, many variations are in use based on the same basic principles.

# Chapter 4

# GENESIS

Some electromagnetic waves such as cosmic rays arrive on earth from the Heavens above but at frequencies generally above our range of interest. The greatest source of electromagnetic radiation is the sun for it is pouring electromagnetic energy into space with a total output of some $3 \times 10^{23}$ kilowatts of which only a tiny portion falls on the earth, yet sufficient to keep us going in both heat and light. Most waves which concern us are generated deliberately by ourselves and here we consider some such waves and how they are produced. Modulation, i.e. the impression of intelligence on the wave, is considered in Section 3.5.

## 4.1 Radio Waves LF to HF

This group of waves is at the lower end of the wave spectrum (Fig.2.6). When the frequency is very low the magnetic field around a wire is usually stronger than the electric field and no appreciable electromagnetic wave is produced. Looked at in another way, it is evident that when a low frequency current flows in a conductor, a magnetic field is set up, however the effect of the field on the distribution of current is generally negligible. As frequency rises the electric field increases until at around say, 15 kHz a wave can be radiated (first investigated by Heinrich Hertz, the German physicist). Radio systems however do not work at frequencies as low as this for many reasons, in fact the "long" waveband does not start until about 150 kHz (2000 metres). The HF band extends to some 30 MHz and in radio transmitters operating within this band, oscillations generated for the purpose of projecting electromagnetic waves into the atmosphere (Sect.2.2) are produced by circuits controlled by inductance + capacitance (tuned) circuits or by a quartz crystal. Electromagnetism abounds in plenty throughout the whole transmitting system for as we have seen any accelerating charge generates an electric field with its associated magnetic field. Nevertheless electromagnetic radiation before the wave reaches the antenna

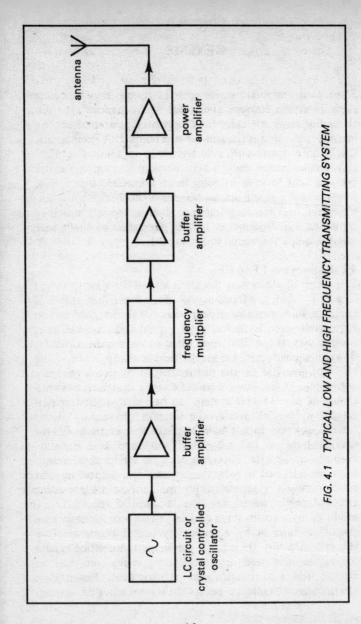

FIG. 4.1   TYPICAL LOW AND HIGH FREQUENCY TRANSMITTING SYSTEM

is kept at a minimum. A simple transmitting system might be as in Figure 4.1.

The oscillator generates the basic frequency, it may be a variable oscillator or crystal controlled. The crystal is a thin slice or plate of quartz. If an oscillatory voltage is applied across two opposite faces of the plate, the latter will vibrate at the same frequency as that of the applied voltage. When this frequency coincides with the mechanical resonance of the plate, the vibration reaches a high amplitude (although still in fact microscopically small) and this can be used to control an oscillator with high frequency stability. So that variations in the load shall not affect the transmitted frequency it is followed by a *buffer* amplifier to isolate the effect of load changes. Depending on the final output frequency, a *frequency multiplier* may be interposed, followed by a second buffer amplifier feeding the main power amplifier. This is one which has sufficient output power to drive the antenna to the degree required.

## 4.2 Radio Waves VHF to SHF

This range covers frequencies from 30 MHz to 30 GHz (see Fig.2.6). Within the range the term *microwave* is likely to arise frequently, it is defined in Section 2.3.3.

Moving up in frequency through the range brings a gradual transition from *lumped* (inductance and capacitance) tuned circuits firstly to those controlled by parallel-resonant (transmission) lines, then above about 500 MHz coaxial resonant lines take over mainly to reduce loss by radiation. Generally lumped circuits are designed around actual components, i.e. one or more capacitors in conjunction with a single inductor. The problem with these is mainly that stray capacitances exist in the circuit, especially within the inductor in which adjacent turns of the winding have capacitances between them. This can be shown to be effectively a shunt capacitance in parallel with the inductor, no great problem at the lower frequencies but as frequency rises the shunting effect increases, giving rise to intolerable circuit losses and/or instability. As an example a stray capacitance of a mere 10 picofarads has a capacitive reactance at 500 MHz of only 32 ohms.

Generally electromagnetic waves over the lower part of this frequency range start off with a quartz crystal for precise control of frequency, followed as necessary by frequency doublers or triplers. The crystal itself may resonate at up to 24 MHz and overtone (harmonic) ones are available up to 200 MHz. Figure 4.1(i) therefore also applies generally to this type of transmitter circuit except that an LC circuit is less likely to be used in the oscillator stage. Again a reminder that the circuit shown is capable of radiating a continuous wave only, modulation of the wave is required for it to carry information.

At the higher frequencies in the range special design techniques become necessary and there is likely to be a complete change in the wave generator employed on moving up through the UHF band. As the frequency of operation approaches 1 GHz, crystal control ceases to be effective, furthermore electronic valves are unable to cope because of electron transit time from cathode to anode. In addition valves have inherent capacitance between the electrodes which precludes very high frequency generation.

Two wave generators frequently encountered for the top of the UHF band and the SHF are the *klystron* and *magnetron*, both of which are found in many different forms and operate from just below 1 GHz to up to 30 GHz or more.

The klystron can be used as a microwave oscillator. The reflex klystron operates mostly at around 2 GHz but is capable of operation up to over 20 GHz. It employs the principle of *velocity modulation* in which modulation of the velocities of the electrons in a beam is used in addition to modulation of the density of electrons as is generally employed at lower frequencies. The method avoids the problems of electron transit time. An *electron gun* is an electrode system within an evacuated tube containing a heated cathode for the supply of electrons plus anodes appropriately biased to form the electrons into a narrow beam. The electrons are then accelerated towards other electrodes in the particular device. In this case the electron beam is velocity modulated in an *interaction gap* as shown in Figure 4.2(i). This forms electron bunches which are returned from the reflector electrode to the interaction gap resulting in oscillation. The complete process is

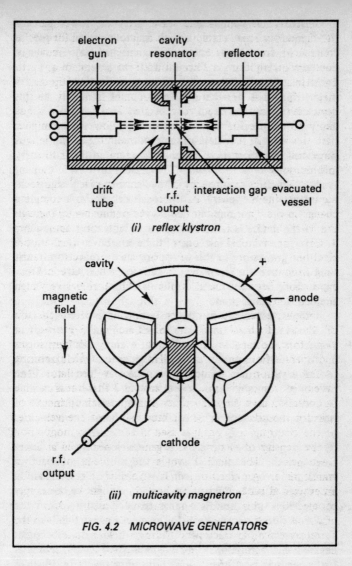

*(i) reflex klystron*

*(ii) multicavity magnetron*

*FIG. 4.2 MICROWAVE GENERATORS*

more than a little complicated.

Magnetrons are employed as microwave oscillators at frequencies from about 1 up to as high as 30 GHz with output powers ranging from a few kilowatts to several megawatts. Early magnetrons date back to around 1939 when the cavity magnetron made possible the development of radar. In this device an electron stream is generated at a central cathode and magnetic fields force the stream to follow a circular path past a series of cavities arranged around the cathode. A simplified representation of the structure of a multicavity magnetron is shown in Figure 4.2(ii). Each cavity is a resonant system which is excited by the electron space charge so converting the d.c. power of the space charge into alternating current power, normally at microwave frequencies. The output of the device is taken through a suitable transmission line or waveguide coupled into one of the resonant cavities.

Other generators in this range are also employed, e.g. the *backward-wave oscillator* which can reach 300 GHz. Semiconductors are also used. Bipolar transistors have been enabled to reach some 5 GHz through special fabrication techniques, especially by the reduction of emitter and base thicknesses so that transit times are minimal. Field-effect transistors are capable of working at even higher frequencies up to 10 GHz or more, again through special manufacturing techniques. Other semiconductor devices have also been developed suitable for operation at these very high frequencies, most relying on a negative resistance characteristic for the production of oscillations. Some examples are:

(i) the transferred-electron oscillator developed by J. B. Gunn, based on the fact that gallium arsenide is a semiconductor which has two conduction bands. Normally electrons are in the lower band but when an applied electric field reaches about 300 kV/m there is a rapid transfer to the upper band in which electron mobility is lower, hence current falls so the current/voltage characteristic turns down, i.e. becomes negative and oscillations can be generated. 300 kV/m may at first appear to be a high voltage but in practice for a device of length, say 0.1 mm, the electric field required to initiate transfer is a mere 30 volts.

(ii)    the tunnel diode which also has a negative resistance section on its current/voltage characteristic, obtained through high doping levels. Under certain bias conditions the depletion layer is narrow and the energy levels are such that the valence and conduction bands overlap. Electrons are then able to "tunnel" across the barrier without the normally required input of energy. It is the tunnelling which results in the negative resistance characteristic and over this the device is usable as a microwave oscillator.

(iii)   the IMPATT diode (Impact Avalanche Transit Time) which can be reverse biased into avalanche breakdown resulting in a negative resistance. The device then functions as an oscillator and an output of up to 100 watts at up to 10 GHz is obtainable.    Above this frequency the device still functions well as an oscillator but generally at lower output powers.

## 4.3  Waves at EHF and Above

The title of this section indicates that we are now concerned with the generation of waves at frequencies exceeding 30 GHz, such waves are not normally considered as radio waves but technically are no different except in frequency. Figure 2.6 indicates that the range includes the visible spectrum bounded by the infra-red and ultraviolet bands, then at even higher frequencies, X-rays, gamma rays and cosmic rays. The last three are beyond the scope of this book but light and the surrounding bands cannot be excluded not only because life depends on vision but also because nowadays these frequencies are capable of transmitting immense quantities of information.

The visible spectrum runs from 380 to 760 nanometres ($7.90 \times 10^{14}$ Hz to $3.95 \times 10^{14}$ Hz — see also Fig.6.5), extending over only one octave yet a frequency band of great importance.  The band contains all the radiant energy which our eyes can utilize, running from red at the lower frequencies to violet at the higher. Devices for generating such waves, so common that we cannot avoid them in everyday life, include tungsten filament lamps, discharge lamps (e.g. fluorescent) and sodium vapour (street) lamps and of course the ubiquitous television tube.  Less frequently encountered are mercury vapour lamps, arc lamps and flash tubes.  More recently light-emitting diodes and lasers have appeared.  Even fire flies and the humble glow-worm are generators of electromagnetic radiation at light frequencies.

# Chapter 5

# FREEDOM

Most radio waves are free in the sense that they are projected into the atmosphere or space and in their travels are neither confined nor guided by artificial means. However Nature has always had her own ideas as to how radio waves should be treated, their so-called freedom is therefore somewhat limited. In a uniform medium waves continue on a straight path at a constant velocity but in a medium in which the characteristics change, they are forced to deviate from a straight-line path and to suffer absorption changes. Although many rules and equations have been developed concerning radio wave propagation, it can hardly be classed as an exact science especially since the rules change with wave frequency. As an example, propagation rules which apply to a low frequency wave have little application at appreciably higher frequencies.

So many variables beset us in getting to grips with radio waves generally that instead of examining the process on a frequency basis as is often done, we consider it more via the transmission path, e.g. ground wave, sky wave and space wave. Ground waves are those which travel along or near the surface of the earth; sky waves travel upwards to ionized layers above the earth (the *ionosphere*) and are returned from there; space waves go straight through the ionosphere into outer space, e.g. to earth satellites.

Whatever the transmission path employed, radio communication requires a transmitting antenna to launch the electromagnetic wave into the atmosphere and at some distance away, a receiving antenna to collect a small sample of it. Antennas are therefore essential components in the transmission of electromagnetic waves through the atmosphere or space so we look at them next although the variations are so great that this can only be done in summary.

## 5.1 Antennas

Electromagnetic waves gain their freedom to skip freely

through the atmosphere or space via an antenna (aerial). Equally an antenna is used to extract a tiny snippet of a wave as it passes. Technically there is little difference between transmitting and receiving antennas working over the same range of frequencies except for size, the transmitting antenna must obviously handle greater electrical powers, accordingly the conductors are larger. On the subject of size it may be of interest to find that the lengths of antennas for transmitting at very low frequencies may be measured in kilometres whereas higher up the frequency range in a portable receiver the length of a ferrite receiving antenna would be no more than a few centimetres.

We touched on antennas in the shape of the dipole radiator in Section 2.2.1, but this was only to show how radiation arises. To assess the performance of an antenna is not straightforward. For a transmitting one, the r.f. power supplied to it is measurable but there are so many factors which affect the radiated signal that no single absolute measurement is possible. Similar considerations apply to the signal output of a receiving antenna. It has therefore become necessary to define a simple elementary antenna against which practical antennas can be compared. The half-wave dipole is one such antenna but possibly in greater favour is the *isotropic*, simply a point source radiating equally in all directions. No such antenna exists but the gains of practical antennas may be rated against it as developed in Appendix 3.

### 5.1.1 Emitting a Wave

We return to the dipole antenna again but now for a better appreciation of how a radio wave is generated. This particular antenna has a simple form and can be imagined as evolving from a quarter-wave transmission line (Chapter 6). Consider a generator to be connected to a pair of wires as shown at (i) in Figure 5.1 and note that at some particular instant when the generator voltage is maximum current flows into the line to charge the capacitance between the two wires. Clearly the current must be zero at the open-circuited end. Alternatively the voltage at this end is at maximum and it is evident that there is a time delay while these effects travel from the generator to the open end. In practice transmission lines are

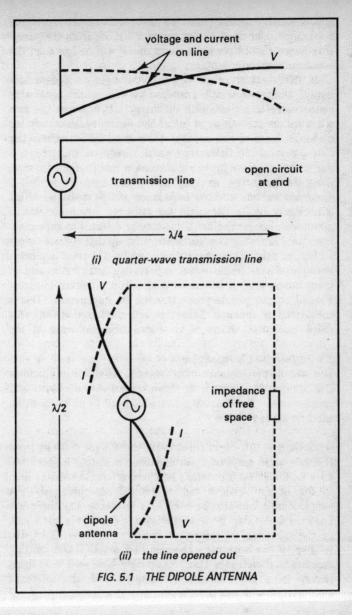

(i)  quarter-wave transmission line

(ii)  the line opened out

FIG. 5.1  THE DIPOLE ANTENNA

not normally opened-ended so there will be some sort of impedance connected across the end remote from the generator hence $I$ will have some value and $V$ will be less than the maximum generator voltage.

If the wires are close together the field created at any instant along one wire is cancelled by that on the other wire hence there is no radiation of energy. If however the two wires are opened out as in (ii) of the figure, because they are no longer close to one another, there is no cancellation of the fields, instead the fields are in such directions that they aid one another. As with the transmission line examined above, the antenna is not working into what might be termed a disconnexion but into the impedance of free space, modified if necessary by an allowance for gases etc. which increase in importance nearer the earth. It may be difficult to appreciate free space knowing that sufficiently far up there where no gas molecules exist at all, radio waves and light abound in plenty. Powerful forces (radio waves and gravity) may be present in space but as long as no matter is present it is still called space. To add to our puzzlement space has its impedance. This is considered in Appendix 2 together with the fact that truly free space does not attenuate an electromagnetic wave at all.

### 5.1.2 Antenna Characteristics
The overall performance of an antenna whatever its shape or size is normally summed up from several measurements and calculations. The characteristics which tell us most about an antenna are as follows:

(i)     *Gain* — this of an antenna has a different meaning from the gain of an amplifier. Antenna gain is simply its performance rated against a generally accepted reference antenna such as the isotropic which can be defined unambiguously. A simple dipole can also be used as a reference and there is a known relationship between the two. The dipole has a gain in the direction of maximum radiation of 1.64 (2.15 dB) relative to the isotropic. The gain of a practical transmitting antenna is therefore a figure expressing how well it radiates energy in a given direction compared with the reference antenna. For a receiving antenna it is a measure of signal pick-up relative to the reference.

(ii) *Directivity* — all practical antennas are to a certain extent directional, frequently they are deliberately designed with this in mind. The directivity of an antenna is most easily displayed on a *polar diagram*. From chosen points on graph paper vectors representing the magnitude of signal transmitted or received and its direction are plotted. The locus of the tips of the vectors forms the directivity pattern (polar

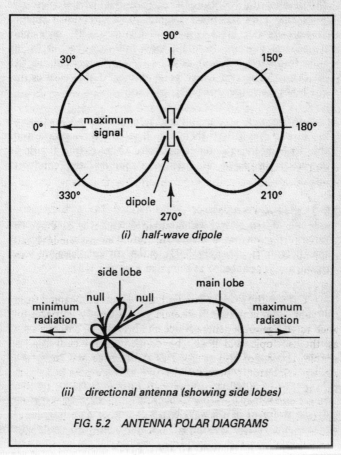

(i) half-wave dipole

(ii) directional antenna (showing side lobes)

FIG. 5.2   ANTENNA POLAR DIAGRAMS

diagram) as illustrated for the horizontal plane in Figure 5.2(i) for a standard half-wave vertical dipole. The response of the dipole at 0° and 180° is maximum and it is said to have a *front-to-back ratio* of 1 since the two responses are equal. At 90° and 270° the response is theoretically zero. Antennas and antenna systems are frequently designed for high front-to-back ratios, i.e. most of the response is in one direction only. Some energy however may be radiated in directions other than that for which the antenna is designed. This is due to diffraction effects (Sect.3.4) which create *side lobes* and a typical polar diagram for a directional antenna showing its side lobes is given in Figure 5.2(ii). Side lobes are separated from the main lobe and from each other by nulls (radiation minima) as shown. They represent inefficiencies in the system hence good antenna design ensures that side lobes are reduced as much as possible.

(iii) *Efficiency* — transmitting antennas do not radiate all the power supplied to them nor is all the wave power available to a receiving antenna delivered to its output terminals. Power is dissipated in antenna structures and conductor resistances.

(iv) *Radiation resistance* — is defined for a transmitting antenna as the power radiated divided by the mean square value of the current at a specified reference point. This latter specification is necessary because the current varies in value throughout the antenna as shown in Figure 5.1(ii).

### 5.1.3 Antennas in Action

In this section we look at a few of the different types of antenna mainly to understand the various methods, shapes and sizes used for launching or collecting an electromagnetic wave. Overall it will be found that there is a strong relationship between antenna size and wavelength.

Generally antennas are divided into two types, resonant and non-resonant. The dimensions of resonant types are of the order of half a wavelength or one of its multiples. In a resonant system the impedance and bandwidth vary when the normal operating frequency is changed as we find with the LC resonant circuit. Non-resonant antennas are less frequency dependent, accordingly are sometimes known as *broadband*.

For maximum power transfer from a feeder to a transmitting antenna or from a receiving antenna to its downlead, there should be a matching of the impedances. This is sometimes difficult to achieve, especially with broadband antennas. The standard dipole impedance at resonance is about 70 ohms, hence a coaxial cable with a similar characteristic provides a good match. For certain applications losses are reduced by stepping up a dipole antenna input impedance for working into a 300 $\Omega$ or 600 $\Omega$ balanced feeder, especially when the latter is long. This results in the *folded dipole* as sketched in Figure 5.3(i). It is in fact a folded full-wave wire resulting in two half-wave dipoles connected in parallel. The folded dipole impedance is about 300 $\Omega$, four times that of the single-wire dipole.

Many types and designs of antennas are required to cover all radio frequencies and operational requirements. The following descriptions of practical antennas is intended as an introduction only. We recall that a radio wave will induce an e.m.f. in any conducting material (even ourselves), however antennas are designed with greater efficiency in mind. There are several references to waveguides, these are explored in greater detail in Chapter 6.

*Long Wire* antennas are used at the longer wavelengths (HF). There are many variations but one which is easily recognized as a half-wave dipole is sketched in Figure 5.3(ii). It is a simple wire horizontal antenna and for example for maximum reception at say 21 MHz requires an overall length of about 20 metres installed at least 25 metres high. To have the down-lead in the centre as shown in the figure may not be convenient hence there are many designs in which the antenna is end-fed. These are appropriately known as *monopole* antennas and they rely on the proximity of the earth for operation, in fact the earth acts as one half of a dipole, although rather inefficiently if its resistance is high. Such antennas are not particularly directional unless a special layout is used as for example in the case of the *rhombic*. This typo is highly directional although at medium frequencies

59

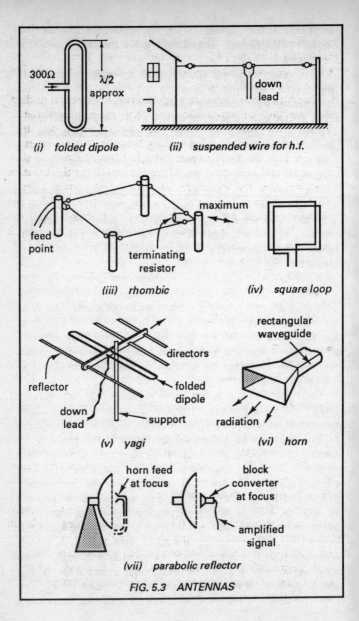

(i) folded dipole    (ii) suspended wire for h.f.

(iii) rhombic    (iv) square loop

(v) yagi    (vi) horn

(vii) parabolic reflector

FIG. 5.3 ANTENNAS

requiring a lot of space. Figure 5.3(iii) shows a rhombic antenna mounted on poles. If designed for say 21 MHz, each side would be over 13 metres long or even longer if higher gain is required. Rhombics are especially used for long distance HF communications.

*Loop* antennas are particularly useful for direction finding. For the sake of explanation the loop antenna depicted in Figure 5.3(iv) is square, circular and other shapes are also used. A vertically polarized wave arriving broadside-on produces no net e.m.f. in the loop because the e.m.f.'s in the vertical sides are equal and in opposition. Turn the loop from this position however, then the wavefront reaches the two vertical sides at different times so giving rise to a phase difference between the voltages induced hence a resultant voltage acting round the loop. This voltage is maximum when the plane of the loop is parallel with the direction of the oncoming wave. This is not a complete direction-finding system because there is a 180° uncertainty as to the direction of the transmitting station. To overcome this a vertical antenna is used in conjunction with the loop; it can be shown that the net signal is increased for one direction but reduced for the 180° opposite direction.

*Rod* antennas are used for shorter wavelengths than those for which long wires are employed. As an example, rod types are invariably used for television receiving antennas working to ground transmitting stations as seen everywhere. These usually consist of a half-wave or folded dipole backed up by directors and (usually) one reflector as sketched in Figure 5.3(v). This is the Yagi-Uda array developed for its good directivity at UHF by two Japanese engineers, H. Yagi and S. Uda. The directors and reflector(s) are simply rods which are excited by the oncoming wave and their spacings from the dipole are such that they re-radiate energy so that it arrives there in phase with the main signal.

*Ferrite Rod* antennas are to be found in portable radio receivers, they are small (typically 12 cm long × 1 cm diameter) and may have long and medium wave coils wound on them. Ferrite is a non-conducting magnetic material of high permeability. The magnetic flux of the wave tends to concentrate through the core and in so doing cuts across the turns of the windings so inducing an e.m.f.

61

*Slot* antennas — theoretically a slot cut in an infinite sheet of metal (the ground plane), if approximately one half-wavelength long acts as a dipole radiator. Practical ground planes are of course smaller, resulting in radiation patterns somewhat different from that of an infinite plane. The slot may be fed from a coaxial line or waveguide, it may even be cut into the waveguide itself. For receiving many slots are cut into a plate and connected together in such a way that the output is the sum of all the individual slots. Frequencies suitable are those above about 300 MHz.

*Horn* antennas are suitable above about 1 GHz and are mainly used for transmitting. At an open end of a waveguide a wave will continue to travel into the open space but because of the sudden discontinuity, much energy is lost. The loss is reduced by terminating the waveguide in a horn to provide a gradual transition into the open air or space. The horn also provides better directivity. Horn shapes may be rectangular or circular, a typical rectangular one is shown in Figure 5.3(vi).

*Reflector* antennas — with satellite television these are now commonplace. They are employed with microwaves (above about 1 GHz), are highly directional and are used for both transmitting and reception. The "dish" (more correctly known as a *focusing pencil beam reflector*) is parabolic in shape and as a receiving antenna serves to direct the electromagnetic wave onto a *feedhorn* coupled either directly to a waveguide or to a frequency changing circuit which reduces the carrier frequency so that the wave may be carried by a coaxial cable. Conversely as a transmitting antenna it can be likened to a searchlight which has a source of light at the focus of the parabolic reflector and projects the light as a parallel beam. Sketches are shown in Figure 5.3(vii), these are the simplest arrangements, others (Cassegrain and Gregorian) incorporate a second, smaller reflector. These are generally used with the large dishes transmitting to satellites.

Above are just a few of the different types of antennas, moreover within each type there are many variations. Only a hint of the constructions in use can be given here.

## 5.2 Ground Waves

Generally ground waves are restricted to low and medium frequencies (Fig.2.6) and are suitable for moderate to long distances such as are required by broadcasting stations. Waves at higher frequencies do not propagate well over the earth because losses increase with frequency, however 3 MHz is by no means the maximum at which a ground wave can be effective. Successful transmissions are used at up to and well over 30 MHz, but the problems of greater attenuation and also of reflections from the ionosphere (see next section) may create interference and signal fading. Fading occurs when the ground and reflected waves arrive at the receiving antenna in opposing phases.

The ionosphere is capable of returning waves back to earth at frequencies up to some 30 MHz so above this the interference problem is reduced. However other factors such as rain, fog and temperature changes begin to make long distance reception difficult. Nevertheless much use is made of the ground wave at frequencies above 30 MHz (i.e. at VHF, UHF and even higher) in systems for which the range is short, perhaps only tens of kilometres, depending on the transmitter power. Typical uses are for television and frequency modulation broadcasting, systems catering for mobile telephones, police and similar services. Thousands of telephone channels are also transmitted over microwave systems but here again distances are relatively short, i.e. generally up to no more than 60 km.

The strength of a ground wave at any particular distance from a transmitter depends on many factors: the transmitted power, wave polarization, frequency, conductivity of the ground, antennas and their heights and the degree to which the curvature of the earth shields one antenna from the other. Altogether this is a formidable collection of variables. There exist simple formulae for calculation of field strength or of distance over which a usable signal will be received. However there are so many variables which cannot be pinned down that such formulae may let us down more often than not. Even highly complicated formulae while doing better, frequently leave much to be desired. Measurement of practical signal strengths is often to be preferred.

The ground wave is most likely to be vertically polarized because a conducting earth would partially short-circuit the electric flux of a horizontally polarized wave (Fig.2.4).

As a ground wave travels away from a transmitting antenna its strength (usually measured in volts, millivolts or micro-volts per metre) is reduced as it spreads out and also as energy is absorbed from it by the ground. A general conclusion is that ground waves at broadcast frequencies travel best over ground or water which has high conductivity and that attenuation of the wave increases with the transmission frequency. Ground losses result from the fact that the electric field of the wave induces currents in the earth which dissipate heat. The lower part of the main wave therefore loses energy which is replaced by diffraction from the portions of the wave above. The velocity of propagation of the wave in the earth is slightly less than that in the air above and this results in a forward tilt of the wave which to a certain extent enables the wave to cling to the earth's surface. The following examples serve to indicate that the order of tilt is quite small. For a wave at 900 kHz travelling over land having a conductivity of 0.015 siemens per metre and permittivity 25, the wavefront will be around $87°$ to the horizontal, i.e. a tilt of $3°$. Over dry soil the tilt increases to around $4°$. For sea water with its higher permittivity (about 80) and conductivity (4 S/m) the angle of tilt at 900 kHz is considerably less at a small fraction of one degree.

A simple formula relating to broadcast transmissions shows that for direct reception of adequate signal levels, the transmission frequency is all important, e.g. if at 100 kHz the transmitter power provides an adequate signal at some 160 km, the reception of a similar signal level at 3 MHz would not be possible beyond a third of this distance.

Much of what has been said in this section ignores the fact that the earth is spherical (almost). Antenna height is there-fore of major importance for as seen from Figure 5.4(i), antenna heights less than those shown cannot result in direct wave transmission unless as is frequently possible, there is some refraction or bending of the wave by the atmosphere as shown in (ii). Such refraction arises from the change in the refractive index of the atmosphere with height (Sect.3.2).

64

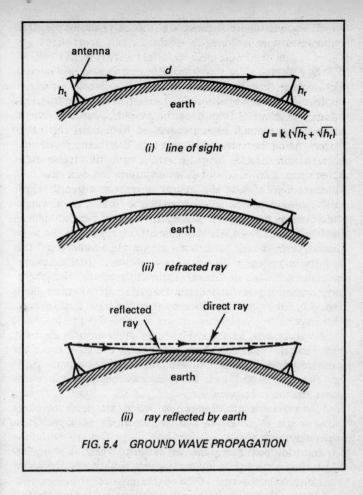

$$d = k(\sqrt{h_t} + \sqrt{h_r})$$

(i) *line of sight*

(ii) *refracted ray*

(ii) *ray reflected by earth*

FIG. 5.4  GROUND WAVE PROPAGATION

However an additional ray path is possible as shown in (iii), arising from reflections from the earth's surface.

Although the effect is not exclusive to this band, radio wave diffraction may be noticeable. Section 3.4 touches upon the phenomenon of diffraction, the effect of which is to allow a radio wave to penetrate the region behind an obstacle. Hills, mountains and buildings create such obstructions, especially when transmitting and receiving antennas are below the radio horizon.

## 5.3 Sky Waves

Considering that generally electromagnetic waves travel in straight lines, a distant radio receiver would be shielded from the transmitter by the curvature of the earth. It looks as though we would have preferred to have a flat earth but Nature had other ideas. However so that homo sapiens is able to communicate by radio over long distances, the ionosphere was added. To this day communication engineers and short-wave enthusiasts still delight in encircling the globe by radio using ionospheric propagation, a technique in which radio waves reaching the ionosphere bounce back as shown in Figure 5.5(i). Clearly A cannot communicate with B directly because the earth is in the way but communication is possible via the reflected wave as shown. The ionosphere is essentially an electrified (ionized) layer of atmosphere surrounding the earth at heights above about 50 km as illustrated in (ii) of the figure (ignore the labelling of the layers at this stage).

An atom with its full complement of electrons is electrically neutral. Take one electron away and the atom becomes positive because it has one more proton than electrons. Add an electron instead and the atom goes negative. In both cases the atom becomes a charged particle known as an *ion* and the process is called *ionization*. When this occurs there are positive and negative ions and free electrons, all available as charge carriers.

The additional energy needed for escape from orbit by an electron is provided in the ionosphere mainly by ultraviolet radiation from the sun. From the energy point of view the radiation is considered to consist of *quanta*, i.e. infinitesimally small "packets" of energy (at light frequencies also known as *photons*). Max Planck (a German physicist) produced a simple formula for calculating quantum energy. From this ultraviolet radiation at, say $10^{16}$ Hz is shown to comprise quanta each of 41 eV energy and at $10^{15}$ Hz, 4.1 eV (Sect.1.1.1).

66

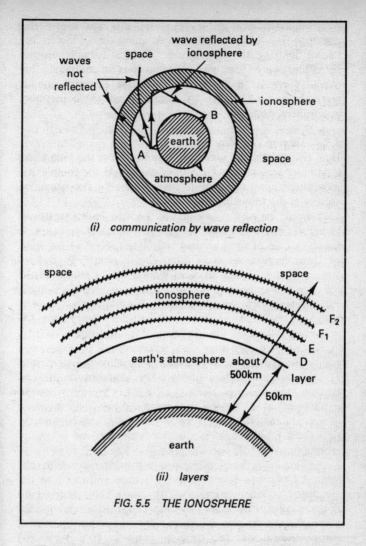

(i)  *communication by wave reflection*

(ii)  *layers*

*FIG. 5.5   THE IONOSPHERE*

The energy which an electron of the oxygen atom requires for release, thereby resulting in ionization of the atom, is just over 13.6 eV and that for nitrogen, 14.5 eV. When a quantum strikes an electron its energy is absorbed by the electron, hence most ultraviolet radiation is capable of freeing electrons from oxygen and nitrogen atoms (or air molecules), i.e. of ionizing the atmosphere. It is a dynamic process, recombination of ions with electrons goes on continually but overall there is an accumulation of free electrons, usually assessed as the number of electrons per cubic metre of atmosphere, symbol $N$.

### 5.3.1 Turning The Wave Back

How electromagnetic waves are returned from the ionosphere is in more senses than one, out of this world for complexity, accordingly we can no more than examine a few simplified aspects of the whole process.

It would be easy to say that as the wave enters regions of higher refractive index it bends over (Sect.3.2) or that it is simply a case of total internal reflection (Sect.3.3), but both of these hypotheses need more explanation. Sir Edward Appleton, the English physicist famous for his studies of the effect of the ionosphere on radio signals, developed a formula for the *complex* refractive index. This relates the magnetic field, the collision frequency of particles, the plasma frequency and of course the frequency of transmission.

*Plasma frequency* may need explanation. A *plasma* is a highly ionized gas with electrons as negative charge carriers as is found in the ionosphere. The electrons vibrate and because they are accelerating charges, they give rise to oscillations of a frequency determined by their number, charge and mass. This is known as the plasma frequency, $f_N$. It is proportional to the charge density and is usually around some $10^9$ Hz but can reach $10^{12}$ Hz or more.

An incoming electromagnetic wave interacts with the ionized layer free electrons and sets them in motion at the frequency of the transmission, the acceleration imposed on them re-radiates the wave in a changed direction. Altogether a complicated process, difficult to disentangle but by making certain assumptions (e.g. that no magnetic field is present) the main formula can be brought down to something more manageable. Thus approximately the refractive index, $n$ at a position in the ionosphere where $N$ is the electron density is:

$$n = \sqrt{1 - (f_N/f)^2} = \sqrt{1 - (kN/f^2)}$$

k is a constant, hence $n$ can be calculated directly from the wave frequency and electron density, the latter can be determined from practical measurements made in the ionosphere.

At the highest ionosphere altitudes the electron density is low because there are only a few gas atoms and molecules around. Lower down these increase and the electron density increases correspondingly. Still further down, that is at the bottom layers of the ionosphere, although gas density is higher still, the ultraviolet radiation from the sun is at its weakest because of absorption and the extraction of energy on its way down through. The result is that electron density falls and as an example it has been estimated that an electron density at a height of 250 km may fall to about one-third of this lower down at 100 km. At the very bottom of the ionosphere the electron density is around zero and from the formula the refractive index = 1. Going up through the ionized layers, $N$ increases, hence $n$ falls.

Snell's Law (Sect.3.2) enables us to see how a wave curves over as it travels upwards in the ionosphere towards the higher electron concentrations. Figure 5.6 shows the effect in all simplicity yet it is sufficient to indicate how a wave is continually refracted (away from the normal because $n$ is decreasing) as it makes its way up through the ionosphere. Ultimately it must run horizontally and here we might imagine that total internal reflection (Sect.3.3) turns it back. Once it has turned downwards a drawing such as Figure 5.6 would show that it is progressively refracted back towards earth ($n$ is increasing therefore the wave is refracted towards the normal). Note that in general terms we say that the wave is reflected even though it is mostly refraction which is involved. This is considered in slightly more detail in Appendix 1 which also includes approximate ranges of electron concentration for the various layers.

What has been considered in this section so far leads to a few general conclusions:

(i)   the angle at which a wave meets the ionosphere is important for clearly at the extreme a vertical wave may

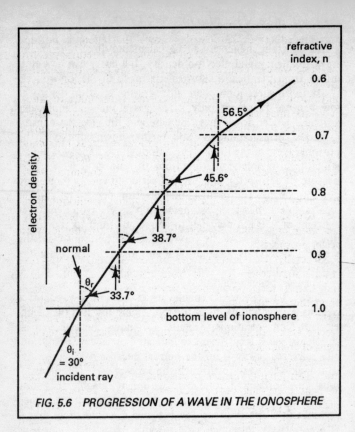

**FIG. 5.6   PROGRESSION OF A WAVE IN THE IONOSPHERE**

not be refracted sufficiently before it has travelled through the layers of greatest charge density;

(ii)   as wave frequency increases $f_N/f^2$ gets smaller, hence $n$ approaches unity so that there is no refraction. This indicates that there is an upper limit to frequencies which are returned to earth;

(iii)   the production of night-time charge densities is obviously less than for day-time because radiation from the sun is reduced.

### 5.3.2 The Restless Layers

The preceding section clearly shows that there is nothing precise about the way in which the ionosphere returns electromagnetic waves to earth. This also follows from the fact that the ionosphere layers vary in their electrical attributes not only from month to month but even from hour to hour. Sunspots, magnetic storms and even some rainstorms affect the degree to which the ionosphere returns radio signals. Such is the variability that expressions such as *maximum usable frequency* (m.u.f.), *optimum usable frequency* (o.u.f.) and *lowest usable frequency* (l.u.f.) are in common use to indicate communication possibilities between any two stations at a given time. Under difficult ionospheric conditions therefore two stations may need to shift frequency appropriately, i.e. from one which is becoming unusable to another with better prospects.

Generally, that is when there is no abnormal interference, m.u.f.'s may reach 20 − 35 MHz around noon with o.u.f.'s and l.u.f.'s several megahertz lower. Earlier and later than noon these drop off considerably, e.g. to a m.u.f. around 10 MHz and l.u.f. around 4 MHz. Remember there is nothing precise about these figures but they do indicate the requirement of frequency shifting.

Exerting a major overall influence on the ionosphere is the sun. This nuclear furnace is surrounded by its *corona*, a halo of light extending outwards by millions of kilometres and ejecting into space streams of particles, mainly of electrons and protons. These atomic particles from the corona stream outwards and are known as the *solar wind* which eventually flows continually around the earth in a way forming a circulating current loop which therefore has an associated magnetic field. Unfortunately for us, Nature in her capricious way of keeping us on our toes, built in a modicum of instability by adding *sunspots*. These are dark patches of gas on the surface of the sun noticeably cooler than the gas around them although at some 4000°C we would consider them to be extremely hot.

Frequently sunspots are created in groups and they are always associated with eruptions of energy and with intense magnetic fields. The normal ionization pattern of the ionosphere is therefore adversely affected by sunspots. These last for varying periods but generally their maximum intensities follow an eleven year cycle on average. It is possible for the magnetic field around a sunspot to collapse, giving rise to an eruption of energy known as a *solar flare*. Such flares eject powerful radiation, mostly of protons and electrons into space, the particles are capable of interfering with the earth's magnetic field and causing blackouts in ionospheric radio communication. Of interest is the fact that whereas light from a flare reaches the earth in a matter of minutes, the charged particles take some $18 - 36$ hours to reach the upper atmosphere, they are of course not electromagnetic waves.

The ionosphere itself is a most complex affair in which the centres of certain layers can be distinguished as shown in Figure 5.5. These are the ones which have the greatest effect on radio transmission. At the bottom, starting at some 50 km high is the D layer. This extends upwards to about 90 km and because of absorption of the sun's rays by the layers above, it is only weakly ionized so its effect is small and because the ionization disperses at night it is only evident within daylight hours with a maximum around noon. The E layer, from about $90 - 160$ km is more strongly ionized than the D layer because there is less absorption of the sun's rays from above. Again maximum ionization occurs around noon and there is little at night, in fact the layer almost disappears.

Above the E layer are the F layers, subdivided into $F_1$ and $F_2$ as shown. Long distance communication relies heavily on these layers but unfortunately they are the most volatile of all since they are more likely to be influenced by particles arriving from the sun. Ionization is the highest of all the layers, even extending through the night. The $F_1$ is the lower of the two layers, its effect at night is small and it has a maximum around noon. The $F_2$, being the uppermost layer has the highest ionization but is less stable. It also varies in height above earth according to season, e.g. during the summer it can reach a height of 500 km but in winter it drops to around 350 km. Maximum ionization occurs in the early afternoon. Ionization of the $F_2$ continues throughout the night, rising rapidly when the sun appears. This layer is especially suitable for long-range short-wave radio communication.

So many variables yet this electromagnetic wave mirror still plays a major role in communications across the world.

### 5.3.3 Sky Wave Propagation

The two preceding sections have outlined the general aspects of the ionosphere, in this section we look at actual radio transmissions in more detail. It is certainly evident that ionization of the various layers differs appreciably. It is the electron density and therefore the degree of ionization together with the transmission frequency which mainly determine the amount of refraction an incident wave experiences (Sect.5.3.1). The amount of refraction influences the *critical angle* at or above which the bending of a wave is insufficient for it to be returned to earth. This is illustrated in Figure 5.7 where wave 1 has not been sufficiently refracted or bent over for its return to earth and therefore it disappears into space. All is not necessarily lost however for communication with

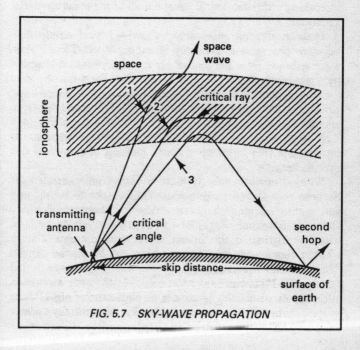

FIG. 5.7   SKY-WAVE PROPAGATION

artificial satellites and space vehicles encircling the earth well above the ionosphere is based on this type of wave.

Wave 2 has been projected upwards at the critical angle, again the degree of refraction is insufficient for a return to earth, this wave is also lost. Wave 3 however meets with more success, continuous refraction as illustrated by Figure 5.6 eventually returns the wave to earth. The *skip distance* shown is that distance measured along the ground from the transmitter to the first point at which the wave is received. A certain area around the transmitter may be served by the ground wave but beyond this there will be a significant area over which no signal at all is received. Waves leaving the transmitting antenna at a smaller angle than that for wave 3 are also returned but with an increase in the skip distance.

Of one thing we are now certain which is that ionospheric transmission is fickle. Happily the modern receiver is fitted with automatic gain control (a.g.c.) which is capable of smoothing out much but at times not all of the signal strength variations.

Even greater distances may be covered by multiple-hop skywave propagation, generally in excess of 4000 km. The wave returned from the ionosphere is reflected upwards again by the ground whereupon it is returned a second time by the ionosphere. In fact several "hops" may occur in succession so that a wave may with favourable conditions, completely encircle the earth by hopping round. It is easy to imagine how waves on a second round trip may interfere with those on the first trip so giving rise to a blurring of the signal or even to "echo".

With so much variability built into the ionosphere it has become necessary for world-wide observatories to be set up for monitoring ionospheric conditions and hence to predict future propagation facilities between working locations. These predictions enable operators to choose optimum transmission frequencies for different times of the day or night.

## 5.4 Satellite Transmissions

Although it is difficult to determine where the sky ends and space begins (or does the sky go on for ever?) it does seem that the explosive increase of satellite transmissions warrants a section on its own. The famous American science writer Arthur C. Clarke first had the idea but it was many years before the rest of us had the equipment. Now (artificial) satellites fly round the earth in profusion. Most are in *geostationary* orbits (stationary with respect to earth), almost circular orbits at over 35 000 km above the Equator. Their speeds are just over 11 000 km/hour and at this height and speed a satellite completes one revolution of its orbit in almost exactly 24 hours (not exactly – this is all very complicated because our year cannot be divided exactly into a whole number of days). Because the earth rotates once in this time, the satellite is always directly over the same point on earth. At this speed and height the centrifugal force on the satellite due to its motion is exactly balanced by the pull of the earth's gravity on it so that it stays in orbit. Satellites not in geostationary orbits are nearer the earth, they travel round it and do not have equatorial orbits.

The geostationary orbit has the distinct advantage that once antennas are sighted onto a particular satellite, no further adjustments are required. On the other hand a non-geostationary orbit implies that the antennas must be constantly realigned to track the satellite – it cannot be slowed or stopped otherwise the centrifugal force would be reduced and the satellite would fall to earth!

Section 5.3.1 shows that there is an upper limit to the frequencies which the ionosphere can return to earth. For satellite transmission therefore the frequencies employed must be above this limit, i.e. they must be able to travel straight through the ionosphere with little or preferably no reflection. This imposes a frequency range in gigahertz and we find that many satellite television transmissions are around 12 GHz. Most satellite systems consist basically of a ground station for transmitting television programmes, data or control signals; the satellite itself and finally a receiving system back on earth. Other systems differ mainly in that the satellite continually transmits information (e.g. weather or military) to earth, signals are only sent up for control purposes. We look to the television satellites for an example. Their function is mainly to receive a signal from earth and retransmit it back over the area required.

### 5.4.1 The Parabolic Antenna

Those of us whose schooldays are long gone may need next to remind ourselves about this geometric shape for it is surely creeping inexorably into our lives as the use of microwaves grows. Proof of this is seen in the pictures of huge so-called "dish" transmitting antennas pointing skywards towards satellites and in the smaller dishes proliferating on the walls and chimneys of our homes for the reception of satellite television.

A parabola is defined as a conic section and it is the geometric curve developed from the simple equation $y^2 = 4ax$. A typical parabolic curve is shown in Figure 5.8. Note that for every positive value of $x$, $y$ has two equal and opposite values. From the figure we see that rays arriving parallel to the principal axis are all reflected to a point in front of the curve known as the *focus*. This is how a receiving antenna functions, for transmitting the arrows are reversed so that power radiated at the focus and directed towards the dish is collimated into a narrow beam (as with a searchlight with its optical mirror).

It is important that for a receiving antenna all rays should arrive at the focus in phase, this condition is satisfied since it can be shown that any ray travels the same distance from a perpendicular line such as AA' to the focus, e.g. ab + bF = cd + dF etc. The same conditions apply in the opposite direction for a transmitting antenna.

### 5.4.2 On The Way Up

Signals are sent up from earth to a satellite from a ground station using a parabolic antenna [see Fig.5.3(vii)]. This type of antenna functions well at the frequencies used and is capable of transmitting most of the signal in a narrow beam aimed accurately at the satellite. There is a small loss owing to diffraction effects (Sect.3.4) around the edge of the reflector which allow energy to "spill over" to behind the dish instead of being reflected forward. The signal is of course an electromagnetic wave in the gigahertz region, it is frequency modulated (Sect.3.5) by the television programme or other information. Obviously for a geostationary system the satellite and ground station are fixed relative to each other and

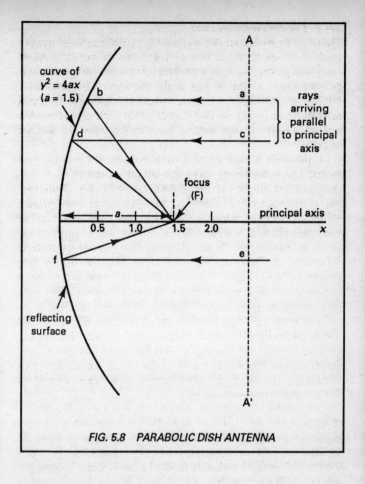

*FIG. 5.8    PARABOLIC DISH ANTENNA*

adjustment to the antenna is only necessary if the satellite digresses at all from its design position (it does happen). Non-geostationary systems require more complicated tracking arrangements. The electromagnetic wave suffers many losses on its way up especially while it is passing through the earth's atmosphere, hence the transmitted power is arranged so that an adequate signal is received by the satellite.

### 5.4.3 Via The Satellite

Continuing with the geostationary television satellite, the programme is received from the ground station on a small parabolic antenna. It is then amplified and changed slightly in frequency. This change is necessary so that the transmitted electromagnetic wave does not enter the receiving system at high level and cause instability. More amplification is then required before the modulated wave, now at a slightly different frequency from that sent up by the ground station, is passed via a waveguide to the transmitting antenna. This is usually a complicated arrangement of paraboloid dishes endeavouring to reduce the coverage area on the ground to that required, the *footprint*. Whereas the uplink to the satellite consists of a pencil-like beam, that returning to earth from the satellite spreads slightly so that on earth the footprint is usually several up to many hundreds of kilometres across, e.g. a single footprint may cover England, Ireland and part of France.

Power for the satellite amplifiers comes solely from sunlight falling on solar cells (Sect.6.5), sometimes spread out on flat wing-like panels or alternatively on drum-shaped satellites, fixed around the outer surface.

### 5.4.4 On The Way Down

In this part of the whole system most compromise has to be made. A sufficiently high level of signal must be provided at the output of the receiving antenna so that a picture clear of noise is obtained. The electromagnetic wave power output from the satellite is limited mainly because the amount of power which can be drawn from the sun is limited. There is plenty of it around but only a small amount can be collected. Output powers therefore are unlikely to exceed 200 watts, generally they are considerably less. On its way down to earth the wave spreads out and experiences loss in travelling through the earth's atmosphere. Eventually it arrives at the receiving antenna which is usually of parabolic shape and must be of sufficient diameter to gather in an ample signal. Within the service area the diameter might be around 60 cm. The parabolic antenna must be aligned accurately with its axis pointing directly to the satellite. The incoming wave is reflected by the antenna onto a *block converter* [see Fig.5.3 (vii)], a device which collects the electromagnetic wave and then changes it to a lower frequency so that it can be fed to the satellite receiver indoors over a flexible coaxial cable instead of a rigid and expensive waveguide.

A practical difficulty with the true parabolic antenna is that the block converter is itself within the path of the free wave and therefore reduces the effective area of the dish slightly. To overcome this some designs focus the wave to a lower point so that the block converter is no longer in the way.

## 5.5 Radar

This is a system which employs electromagnetic waves for direction and range finding. It uses a beam of waves to detect targets such as ships and aircraft and it is also extensively used in meteorology. Waves are focused into narrow beams at frequencies upwards from just below 1 GHz (wavelengths in the cm range). Energy is re-radiated by metal objects in their path. In Appendix 2 it is shown that free space has an impedance: the atmosphere also has an impedance and as might be expected, not very different from that of space. Accordingly the atmosphere can be considered as a transmission line and when it carries a beamed electromagnetic wave, as shown in Section 6.1.1 any object in the path of the wave which does not have the same characteristic impedance as the atmosphere, results in reflection of some of the wave energy. This is picked up at the radar station and the direction and distance away of the object can be determined.

There are many different arrangements of radar systems but probably most are of the *monostatic* type in that a single antenna is used for transmitting the searching signal and also for reception of echoes. Although some systems employ a continuous searching signal, most use short duration pulses, such a system is described in the following section.

It all looks straightforward but the process is beset with difficulties, perhaps the most obvious is that the re-radiated wave may be very weak (e.g. from a small aircraft head or tail on) and it could be lost in the system noise. Rain and clouds may also give rise to more attenuation than we would like.

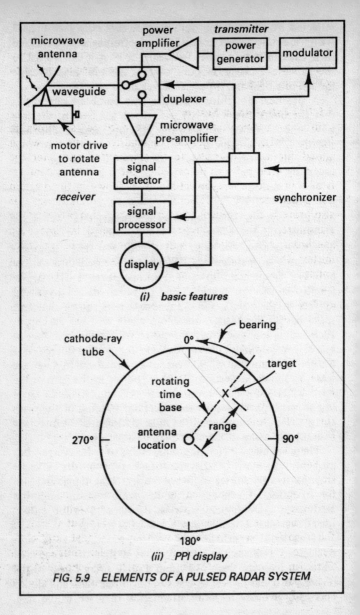

(i) basic features

(ii) PPI display

FIG. 5.9  ELEMENTS OF A PULSED RADAR SYSTEM

It is this which had led to pulsed systems which allow considerably higher peak powers to be transmitted during the short duration of a pulse compared with the limited power which can be generated at these very high frequencies with continuous wave systems.

### 5.5.1 A Pulsed Radar System

Central to a monostatic system is a *duplexer* as shown in Figure 5.9(i). This is basically a send/receive switch which allows the outgoing pulse to be transmitted and later the incoming echo pulse to be passed to the receiving system. It is of course not a mechanical switch as shown in the diagram but a considerably more sophisticated electronic device. It also protects the receiving system from the high power of the transmitter. The whole system is controlled in time by a *synchronizer*. This first activates the modulator of the transmitter which through the power generator applies a high voltage pulse to the power amplifier, usually either a klystron or for higher powers a magnetron (Sect.4.2). The duplexer connects this high power microwave pulse to the antenna.

The incoming reflected signal is switched by the duplexer to the microwave pre-amplifier for initial amplification. The signal is detected and then processed for rejection of unwanted signals. The output of the signal processor is then fed to the display system, usually of the *plan-position indication* type as frequently seen on films and television. The antenna is rotating continually in its search for objects and the time base of the display cathode-ray tube rotates in synchronism. The bearing and range of any object picked up by the radar is indicated on the screen as sketched in Figure 5.9(ii).

Speeds of moving objects can also be measured but for this continuous waves are used. The echo is detected and its frequency is compared with that of the transmitted frequency. If the target is stationary the echo will of course be at the same frequency as that transmitted but if moving the echo will leave the target with an additional positive or negative speed imposed on it (the Doppler shift — after Christian Doppler, an Austrian physicist). The echo frequency is compared with the transmitted frequency and from this the speed of the target is calculated.

# Chapter 6

# CONSTRAINT

In Chapter 5 we develop an overall appreciation of the radio wave, how it can be left to its own devices or alternatively guided in some way. One might argue that when a wave is concentrated into a beam, this is hardly freedom as the title of the chapter suggests but the term is meant to indicate that the wave is free to travel unconfined by things material. Conversely in this chapter we direct our attention mainly to electromagnetic waves enclosed within some sort of guide such as in the transmission line, waveguide and glass fibre. The latter also happens to be a waveguide but may not generally be thought of as such. Many of the theoretical concepts relating to transmission lines are also applicable to waveguides and we must remember always that waves travelling along a transmission line or within a waveguide are fundamentally the same as the free-space electromagnetic waves of Chapter 5.

## 6.1 Transmission Lines

These are systems of continuous conductors of constant cross-sectional dimensions throughout their length used to carry electromagnetic waves between two fixed points. A single wire on poles with earth return is such a line, better still a pair of wires running parallel either overhead or within an underground cable. The coaxial cable is also a transmission line and in this case the outer conductor (the sheath) encloses the waves. An idea of the construction of a coaxial cable is given in Figure 6.1(i). The cable conductors consist of a flexible copper tube or braiding through the centre of which runs a single copper wire. This central wire is held in position by polyethylene support discs or by a continuous cellular polythene filling. In both cases the insulation between the two conductors is mainly air.

Coaxial cable is especially useful at the higher frequencies (above about 10 MHz) compared with a pair of wires in the open because the electric and magnetic fields of the signal

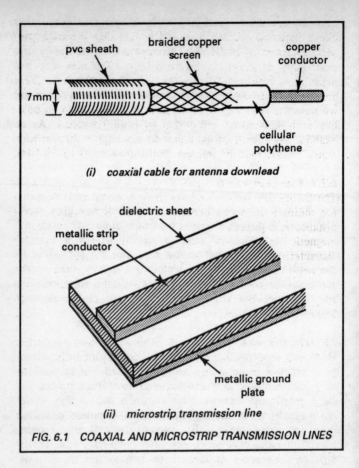

(i) coaxial cable for antenna download

(ii) microstrip transmission line

FIG. 6.1 COAXIAL AND MICROSTRIP TRANSMISSION LINES

terminate on the inner surface of the outer conductor. Coaxial cables are therefore completely enclosed, there is effectively no signal radiation nor is the cable affected by external interference.

*Strip transmission lines* come into their own in the gigahertz range provided that only short runs are required and the transmitted power is low. Typically this type of line consists of a copper conductor running between two *ground planes*, the latter most conveniently being formed from copper-clad polyethylene sheet. Such lines can be imagined as being rather like a flattened coaxial cable and in fact their electrical properties have many similarities. *Microstrip* lines use one ground plane only as shown in Figure 6.1(ii). The dielectric is a material of high permittivity (Sect.1.1.2) and low loss. These are a miniaturized type designed especially for integration and we note the similarity with the overhead single wire on poles but with dimensions more than a little reduced. As an example a 10 mm strip on a ground plane 25 – 30 mm wide would be suitable for frequencies from about 5 to 10 GHz.

### 6.1.1 Line Coefficients

The theory of transmission lines is more complicated than that for ordinary networks because the wave is travelling over a distance and therefore time is involved. How fast an electromagnetic wave moves along a line depends on the line characteristics and it will not be as high as $3 \times 10^8$ m/s as for the radio wave because transmission is not in free space. The full range of transmission line characteristics is derived from the four *primary coefficients*, these are quoted per unit length, usually the metre:

(i) $R$, the *series resistance* in ohms of the two conductors as a loop. Skin effect (i.e. the tendency for the current to flow more on the surface or "skin" of a conductor as frequency rises) causes the value of $R$ also to rise with frequency.

(ii) $G$, the *shunt conductance* in siemens arising from leakage currents flowing through the dielectric material separating the conductors.

(iii) $L$, the *series inductance* in henries for the two-wire loop (note that even a single length of wire has inductance because when it carries an alternating current it is surrounded by a changing magnetic field).

(iv) $C$, the *shunt capacitance* in farads is simply that between the two conductors.

The primary coefficients, all of which can be measured lead directly to calculation of the *secondary coefficients* from which the electrical conditions under which the line will

85

will operate can be calculated. These are:

(i) *Characteristic impedance* — theoretically this is the impedance measured at one end of a transmission line which is infinitely long on the basis that a wave applied to the sending end will travel on and on and never result in the reflection of energy back. Although the concept of an infinitely long line is purely theoretical, such a definition does not preclude practical use for it can be shown that a length of line terminated by its characteristic impedance behaves as though the line were infinitely long. In practice the characteristic impedance can be calculated from a knowledge of the primary coefficients.

If a practical line is terminated in an impedance other than its characteristic impedance ($Z_0$), then *reflection* arises. Consider the extreme case when the distant end of the line is on open circuit. An electromagnetic wave arriving at the disconnection has to go somewhere and the only course open to it is to travel back along the line towards the sending end. By so doing there are now two waves at the same frequency on the line travelling in opposite directions. This results in *standing waves* for which the electrical level at any point does not vary and points of electrical minimum level (*nodes*) and maximum level (*antinodes*) are not propagated along the line even though the waves creating them are.

Calculation of the characteristic impedance of a practical transmission line is easily accomplished once the four primary coefficients have been measured or calculated. Also, provided that cooperation at the distant end is available to open-circuit the line and then short-circuit it, impedance measurements can be made for each condition at the testing end and it can be shown that the characteristic impedance is given by the square root of the product of these two measurements.

Two well-known formulae link characteristic impedance with line dimensions. For a two-wire transmission line well above the earth and with air as the dielectric the characteristic impedance is given approximately by:

$$Z_0 = 276 \log D/d$$

where the wire centres are $D$ apart and each has a diameter $d$.

For a coaxial line:

$$Z_0 = 138 \log B/A$$

where $B$ represents the internal diameter of the outer conductor (screen) and $A$ represents the outer diameter of the inner conductor.

If a dielectric other than air is used between the conductors the result is multiplied by $1/\sqrt{\epsilon_r}$ where $\epsilon_r$ is the relative permittivity of the material (Sect.1.1.2).

Generally characteristic impedances of coaxial cables are of the order of $50 - 100$ ohms. As an example, when $B/A = 3.2$, $Z_0 = 69.7$ ohms. The type of cable illustrated in Figure 6.1(i) has a characteristic impedance of around 75 ohms.

Reflections on a line and the associated power losses must be avoided or at least reduced to a minimum by terminating the line at both ends with its characteristic impedance.

(ii) *Propagation Coefficient* — this when calculated for a particular line expresses the line attenuation (the *attenuation coefficient*, usually in decibels per metre or kilometre) and the phase change which a signal undergoes as it travels along the line (the *phase-change coefficient*, usually in radians per metre). Again this is calculated from a knowledge of the four primary coefficients.

(iii) *Velocity of propagation* is a by-product from calculation of the propagation coefficient for if the value of the phase-change coefficient ($\beta$) is known, say in radians per metre, and the wave frequency ($\omega$) is in radians per second ($\omega = 2\pi \times$ frequency) then the velocity of propagation is simply $\omega/\beta$ in metres per second. As an example, for a practical 0.91 mm diameter paper-insulated trunk telephone cable at 100 kHz, $\beta$ might be $3 \times 10^{-3}$ radians (about 0.2 degrees) per metre. The wave frequency in radians per second is $2\pi \times 10^5$ hence:

$$\text{velocity} = \omega/\beta = \frac{2\pi \times 10^5}{3 \times 10^{-3}} = 2.09 \times 10^8 \text{ m/s}$$

which we note is considerably lower than the velocity of electromagnetic waves in free space.

## 6.2 Waveguides

Electromagnetic waves may also be confined and travel within a rigid metal guide, usually of rectangular or circular cross-section. The waveguide differs from the coaxial cable mainly in that its dimensions are closely related to the wavelength being carried, it is rigid and requires no centre conductor. Waveguides take over from coaxial cables when the wavelength is less than about 10 cm ($f = 3$ GHz). A sketch of a rectangular guide is shown in Figure 6.2(i). Usually the dielectric is dry air. As an example, for internal dimensions of $2 \times 1$ cm the guide would carry frequencies running from about $10 - 15$ GHz ($\lambda = 3 - 2$ cm). For considerably higher frequencies, say $60 - 90$ GHz the guide internal dimensions might be $0.3 \times 0.15$ cm and we note that in these two cases the longer internal dimension is double that of the shorter which is often the case. From the figure it is obvious that waveguides are expensive especially since they must be machined accurately and be waterproof, they are therefore mainly used on short runs, e.g. feeds to ground or satellite transmitting antennas and feeds from generators into microwave ovens.

An electromagnetic wave travelling within a waveguide does so by successive internal reflections from the guide walls, effectively it zig-zags along the guide. This is because the normal wave cannot travel straight through with the electric field running parallel to the walls of the guide because the walls are highly conductive and therefore would short-circuit the field. The electric lines of force must therefore always be at an angle to the guide walls. A wave can therefore be considered to travel along the guide as indicated in Figure 6.2(ii). This sketch is of course an extremely simplistic way of looking at a wave but its purpose is solely to indicate the mean direction in which a wave moves. The magnetic field creates surface currents in the guide walls through skin effect so giving rise to a small waveguide loss, say of the order of 0.18 dB per metre at 10 GHz. We will see later that this compares unfavourably with glass fibres for which the attenuation may

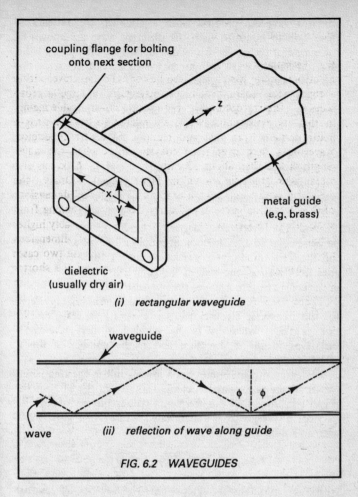

coupling flange for bolting
onto next section

z

x
y

metal guide
(e.g. brass)

dielectric
(usually dry air)

*(i)   rectangular waveguide*

waveguide

$\phi$ | $\phi$

wave     *(ii)   reflection of wave along guide*

### FIG. 6.2   WAVEGUIDES

be only a few decibels per kilometre and that at considerably
higher frequencies.

### 6.2.1 Propagation Modes

Figure 6.2(ii) shows that transmission of an electromagnetic
wave along a waveguide is entirely through the dielectric and
not along the guide walls. There is more than one way in

89

which a wave can progress along a guide through reflections from wall to opposite wall, the different ways are known as the *propagation modes* for which details of only one of the electromagnetic wave components need be quoted since the other is always at right angles.

The simplest or fundamental rectangular waveguide mode is designated $TE_{10}$, the letters standing for *Transverse Electric* to show that there is no electric component in the direction of transmission [the $z$ direction – see Fig.6.2(i)]. The subscript figures indicate the number of maxima (half-wavelengths) occurring in the electric field in the $x$ and $y$ directions respectively – somewhat difficult to visualize perhaps but imagining such a wave in the opening of the guide in Figure 6.2(i) could help. Of most importance are the TE and TM (*Transverse Magnetic*) modes which are always perpendicular to the direction of propagation. As an example, for the $TE_{10}$ mode, at a particular instant the electric field goes from zero at one wall of the guide, through a maximum in the centre to zero again at the opposite wall.

There are many possible transmission modes, especially for the transverse electric field pattern. Its general designation is $TE_{mn}$ where m is the number of half-wavelength variations in the $x$ direction and n the number in the $y$ direction.

Each transmission mode has a lower limit to the frequency which can be propagated. As an example, for the $TE_{10}$ mode it can be shown that this frequency (the frequency of cut-off, $f_{C0}$) for a waveguide of dimensions as shown in Figure 6.2(i) with an air dielectric:

$$f_{C0} = \frac{c}{2} \sqrt{(m/x)^2 + (n/y)^2}$$

with the reminder that $c$ represents the velocity of electromagnetic waves in space. For any other dielectric its permeability and permittivity must also be taken into account so the formula becomes rather more complicated.

For the $TE_{10}$ mode m = 1, n = 0, hence with an air dielectric:

$$f_{C0} = c/2x$$

hence the cut-off wavelength $= 2x$ and in practical waveguides this is approximately so, e.g. the calculated lowest frequency for a practical $7.05 - 10$ GHz waveguide of internal dimensions $x = 2.85$ cm, $y = 1.26$ cm is $c/2x = 5.26$ GHz. The $TE_{10}$ mode has the lowest cut-off frequency of all modes.

### 6.2.2 Input and Output

Launching a wave into a guide requires some means of transforming the radio frequency energy flowing into the feeder into electromagnetic wave energy. This assumes that the generator itself is not mounted within the guide as it may well be at the higher frequencies. A simple method is by extending the centre conductor of a coaxial cable carrying the electromagnetic wave into the end of the guide. In a way it then acts as an antenna. Typically the centre conductor or *probe* may penetrate into the guide through a side of the larger dimension and clearly the electric field generated in the guide would normally propagate in all directions. To obviate this a short-circuit is placed behind the probe and energy is reflected from the short-circuit so that it aids the wave propagating in the forward direction. The desired mode of transmission is obtained through the method of alignment of the probe within the guide.

Instead of a probe for setting up the electric field, a current-carrying coil may be inserted within the guide so that the incoming wave induces the magnetic field. Such feeds are reciprocal devices so can also be used for extracting an electromagnetic wave from the guide at the receiving end.

### 6.3 Fibre Optics

So far we have considered constrained or guided communication systems in the form of two-wire lines, coaxial cables and waveguides. We now jump several orders of frequency from the EHF band (Fig.2.6) to the visible spectrum to consider a further, more recently developed system which has untold advantages and is rapidly replacing many of the existing older methods. This is based on the glass or plastic *fibre*. Section 6.2 indicates that there is a relationship between the dimensions of a waveguide and the wavelength of the electromagnetic wave carried, the fibre is no exception for it is a waveguide but now designed to carry light waves which have such tiny wavelengths compared with those we have studied so far that the fibre is extremely small. To the human observer it is a waveguide which is hardly visible, frequently being a small fraction of one millimetre in diameter, i.e. somewhere around the thickness of a single human hair. Yet the information flow along such a fibre when compared with the systems mentioned above is truly phenomenal. Little imagination is therefore required to appreciate the benefits of for example a few such fibres in undersea cables compared with groups of coaxial lines.

Stripped of all accoutrements, a fibre-optic transmission system consists of (i) a modulated light source, e.g. a light emitting diode (LED) or laser; (ii) the length of fibre connecting the sending and receiving ends with line repeaters or regenerators as required; and (iii) a receiver or photo-detector to demodulate the electromagnetic wave at the far end. The two directions of transmission originally required two fibres but more recent developments now enable both directions of transmission to be accommodated on a single fibre.

In slightly more detail the system is as shown in Figure 6.3. Couplers and connectors are very important in the system. Whereas in a normal metallic system connections are straightforward, usually a simple soldering job, in fibre systems the problems of alignment and the losses incurred where this is not perfect have needed much research and trial. Unless care is taken, coupling light from a source into a fibre can be very inefficient because effectively the light has to be "squeezed" down to fibre size and misalignment with a hair-breadth fibre can result in excessive light loss. On the other hand at the receiving end light emanates from the fibre directly onto the detector surface which is of greater area, losses are therefore low.

Whereas so far we have been describing electromagnetic waves mainly in terms of their frequencies, by continuing to do so for the optical range would involve us in inordinately large numbers. Generally therefore scientists prefer to discuss such waves in terms of their wavelengths, usually in micrometres ($10^{-6}$ m) or nanometres ($10^{-9}$ m).

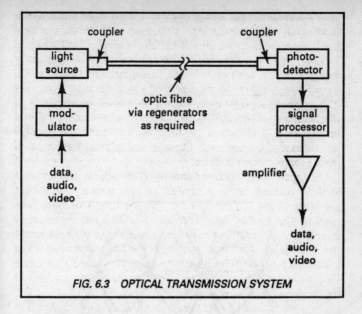

FIG. 6.3 OPTICAL TRANSMISSION SYSTEM

### 6.3.1 The Fibre

A single fibre can be looked upon as a dielectric waveguide and therefore it will transmit electromagnetic wave energy. The main problem however is the leakage of energy from the outer surface of the fibre into whatever surrounds it. This is successfully overcome for both optical and infra-red waves by coating the fibre with a cladding of another glass or plastic material but one which has a slightly lower refractive index.

We met total internal reflection in Section 3.3 where it is shown that under certain conditions a ray of light striking the junction between two translucent materials having different refractive indices can be completely reflected. This is illustrated by Figure 3.2 and if it were not for this particular phenomenon, fibre-optic transmission as we now know it would not be with us today. With a single hair-like glass or plastic fibre therefore, provided that a ray of light travelling within the fibre is moving in such a direction that it strikes the surface of the fibre at an angle of $\theta_i$ to the normal which is equal to or greater than the critical angle, then the ray is

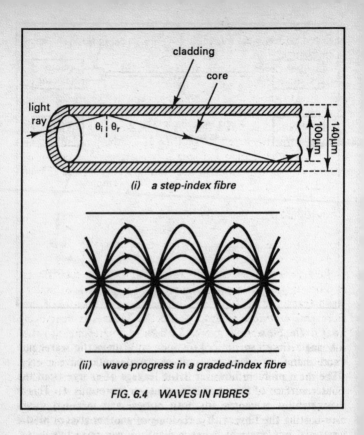

(i)  *a step-index fibre*

(ii)  *wave progress in a graded-index fibre*

FIG. 6.4   WAVES IN FIBRES

reflected from the surface as shown in Figure 6.4(i) with an angle of reflection, $\theta_r$ equal to $\theta_i$. Typical dimensions are shown to the right of the drawing. Core diameters of 50 and 200 $\mu$m are also in common use. The drawing shows that as the ray travels onwards it is again reflected in the same way. The rays must therefore be at or beyond the critical angle for the particular medium for them to be transmitted onwards without loss. This is a *step-index* fibre, the term refers to all fibres which consist of a central core surrounded by a cladding, both core and cladding having their own constant values of refractive index throughout.

Typically a glass fibre with a refractive index of 1.45 enclosed within a glass cladding of refractive index 1.43 has a critical angle of $\sin^{-1} 1.43/1.45 = 80.5°$.

All-glass fibres have the lowest transmission losses. Conversely all-plastic fibres have the highest losses, this restricts their use to short runs ($< 100$ m). Core diameters for these may be 1 mm or more.

In *graded-index fibres* there is no single step from material of one refractive index to another of a different index, the refractive index varies with the distance from the centre of the fibre. An approximate relationship describes the change of index:

fractional refractive index change,

$$\Delta = (n_1 - n_2)/n_1 \qquad (n_1 > n_2)$$

Progression of an electromagnetic wave along such fibres is more than a little complicated but we can at least understand that a ray of light travelling out from the centre of the fibre and experiencing lower refractive indexes is bent over away from the normal just as for a wave progressing into the ionosphere as shown in Figure 5.6. Ultimately the ray angle exceeds the critical angle for the particular material and it is therefore totally reflected. Thereon it is bent towards the fibre centre as it is now experiencing an increasing refractive index (i.e. it is continually bent towards the normal). On crossing the fibre axis the process repeats and the wave oscillates to and fro as it propagates along the fibre. This is illustrated in Figure 6.4(ii), never forgetting that we are rather inadequately trying to picture the highly complex electromagnetic wave as a few lines.

Generally but not exclusively fibre systems work within the range 0.5 to 1.6 $\mu$m as illustrated by Figure 6.5, this range is seen to include much of the visible spectrum extending well down into the infra-red. In fact glass does not transmit light as well in the visible region as it does in the upper part of the infra-red because attenuation is greater. As wavelength is reduced below 0.5 $\mu$m, i.e. in the blue, violet to ultraviolet region, the attenuation becomes even greater. The efficiency

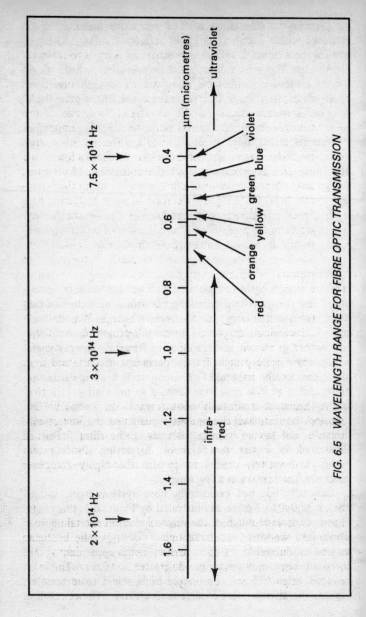

FIG. 6.5 WAVELENGTH RANGE FOR FIBRE OPTIC TRANSMISSION

of glass is at its best around 0.85 $\mu$m and also from 1.1 to 1.6 $\mu$m.

The total loss of a fibre arises mainly from three effects:

(i) *absorption* — there is a certain amount of molecular absorption which is a natural feature of the glass itself, this mainly occurs in the ultraviolet so is one of the factors limiting the useful range to 0.5 $\mu$m. Impurities also create loss, these are metals which absorb energy from the wave when moving electrons from a lower to a higher energy state through the absorption of photons;

(ii) *scattering* — arises from random variations in the refractive index of the glass, resulting in some scattering of a beam of light passing through. It can be shown that scattering is proportional to $\lambda^{-4}$, showing that as wavelength increases scattering loss decreases considerably hence providing an incentive to work in the infra-red region (Fig.6.5);

(iii) *bending loss* — clearly when a fibre is bent a ray entering the curve may strike the interface with the cladding (step-index fibre) at an angle which is less than the critical angle and total internal reflection does not therefore occur but something less. Generally losses are not appreciable provided that the radius of curvature does not exceed about 10 cm.

As enumerated above, the losses may look formidable but research has continued apace with reduction in fibre attenuation as the key to success. Whereas the earliest fibres had attenuations at the very best of 20 dB/km, more recent ones are down to a small fraction of one decibel per km over much of the useful wavelength range.

### 6.3.2 Light Sources

Slowly but surely digital transmission systems are taking over from the well-tried analogue ones. In these the baseband signal modulates the optic carrier directly according to the well known principles of modulation (Sect.3.5). The take-over by digital is not completely of course for we ourselves are analogue talkers and it does not seem that we are ever likely to become digital, therefore at least those parts of a communication system which involve human beings must remain analogue. For the benefit of some readers it may therefore be appropriate here to mention briefly the essential features of digital transmission.

The system used almost exclusively is known as *binary* (of two). Information is transmitted as a series of pulses and spaces. This is rather like an updated Morse Code except that instead of dots and dashes we use signals generally described as 1's and 0's, each known as a binary digit (bit). As an example a letter might be coded in binary as 1000101 and when transmitted a perfect signal would appear on an oscilloscope as shown in Figure 6.6. Analogue waveforms are converted to digital by *sampling* which involves measuring the level of the waveform at pre-determined intervals, then assigning a binary code according to the level of voltage measured.

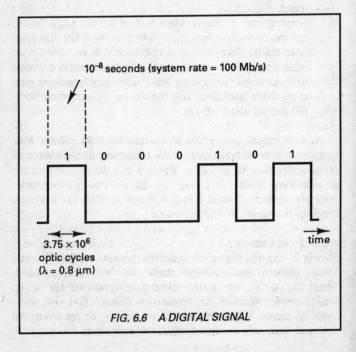

$10^{-8}$ seconds (system rate = 100 Mb/s)

1  0  0  0  1  0  1

$3.75 \times 10^6$
optic cycles
($\lambda = 0.8\ \mu m$)

time

FIG. 6.6   A DIGITAL SIGNAL

At a system rate of, for example 100 Mb/s ($10^8$ bits per second) a single pulse (or no pulse) duration is $10^{-8}$ seconds as shown on the figure. If the wavelength, $\lambda$ of the transmission is, say 0.8 $\mu$m, then taking $c$ as the velocity of the wave, $T_1$ the time of one optic cycle is $2.667 \times 10^{-15}$ seconds. Hence during a pulse duration of a mere $10^{-8}$ seconds there are $10^{-8} / 2.667 \times 10^{-15} = 3.75 \times 10^6$ optic cycles, i.e. well over 3 million cycles of the electromagnetic wave for a $10^{-8}$ second pulse, again reminding us of the almost unbelievable frequencies of light.

Since the transmitted waveform has two states only, impressing a digital signal onto a carrier simply involves switching the carrier on or off at the appropriate times. System rates vary from less than 100 kb/s to as high as 1 Gb/s ($10^9$ b/s) or more which as one might expect is a high rate and somewhat expensive.

Here the term *light* also includes that part of the infra-red shown on Figure 6.5. The first requirement of any light source employed for optical communications is that it has an emission wavelength coinciding with the low loss section of the characteristic of the fibre with which it is to be used. It must also be capable of operating at high bit rates meaning that its light output reaches maximum or ceases almost instantaneously when the input is switched on or off. Preferably also the output light power should be linear with the magnitude of the input signal. Most suitable as light sources are the light-emitting diode (LED) and the semiconductor laser, they are both also conveniently small devices and consume only moderate power.

Generally lasers are used in wide-bandwidth, long-range communication systems mainly because of their fast response to input signals. LED's are an alternative for narrow-bandwidth, shorter-range applications. Both devices operate through the transition process involved when electrons change between the valence and conduction energy bands of certain semiconductor materials. In these materials electrons which normally would be in the outermost orbits of the atoms (*valence electrons*), when provided with sufficient energy are able to escape from orbit and hence be free to act as current carriers, generally known as *conduction electrons*.

The difference between the valence and conduction energy levels ($E_g$) is known as the *band-gap* and for an electron to be raised to the conduction level it needs to receive sufficient energy (i.e. in excess of $E_g$) in the form of a *photon* which we might briefly describe as a "packet" of light energy (see Appendix 5). On the other hand an electron so excited may be encouraged to give up this energy in the form of light and in so doing return to the valence band. Now if a sufficiently large number of electrons can be excited to the conduction band, and then be induced to return en masse to the valence band, photons with energies just larger than the band-gap must be released but of course we are not talking in terms of a few photons but in fact millions and millions of them so releasing energy in the form of a flash of radiation. This optical energy is launched directly onto the fibre via a coupler. Max Planck (a German physicist) gave us a basic relationship between energy and waves and from this we can calculate the wavelength of the radiation produced:

$$\lambda = hc/E_g \text{ metres}$$

where $h$ is Planck's constant ($6.626 \times 10^{-34}$ joule-seconds) and $E_g$ is the band-gap energy in joules.

If we calculate $\lambda$ directly in micrometres with $E_g$ in electron-volts (for electron-volts see Appendix 4 if required):

$$\lambda = 1.24/E_g$$

Suppose the energy gap $E_g$ is 2 eV, then:

$$\lambda = 1.24/2 = 0.62 \,\mu m$$

which we note from Figure 6.5 is orange light.

From this it is evident that the emitted wavelength depends on the band-gap energy for the material employed as the light source. As a practical example, the band-gap energy for a light-emitting semiconductor of gallium arsenide is 1.4 eV giving a wavelength of $0.89 \,\mu m$.

Whether analogue or digital modulation is employed depends mainly on the type of information to be carried. If the input is already digital, then generally it remains so over the fibre. If analogue (e.g. video signals in television links) the choice depends on several factors, especially on distance, for example for short runs video signals are likely to remain in analogue form.

Taking an LED as an example, suppose it has an input current of $i$ amperes. This is equivalent to an input of $i/e$ electrons per second where $e$ is the charge on each electron $(1.602 \times 10^{-19}$ coulombs). Now only a fraction of these electrons will be effective in changing between the conduction and valence bands to produce photons as shown above, call this fraction $\eta$. If the band-gap energy $E_g$ is in electron-volts, then the optic power output:

$$P = \eta i E_g .$$

Note that this is the power output of the device, the power actually available inside the fibre is less because of coupling losses.

So far we have concentrated on the LED as the optical light source. This turns on whenever a positive current flows through it. In contrast laser diodes are biased so that they do not radiate unless a current greater than a threshold value is applied. In a digital system an incoming binary digit 1 drives the current beyond threshold and a flash of light is emitted. An incoming 0 does not drive the device beyond threshold and so it does not operate. The laser has a narrower output spectrum and can be modulated at higher rates compared with the LED.

### 6.3.3 Light Detectors
On arrival at the distant end of a fibre system circuitry for the detection or demodulation of the incoming signal must be provided. Its function is to convert the optic signals to electrical. The devices employed must therefore be able to respond quickly, bearing in mind that many system rates are in Mb/s and may even be as high as Gb/s. The latter systems however are complex and expensive. A light detector must be capable of changing its state rapidly. This is indicated by the *rise-time* which shows the time the output current takes to change from 10% to 90% of its final value when the optic

power input is a step, i.e. it changes from low to high in zero time. Equally the output current must fall rapidly when the optic power input ceases.

A second important detector property is its *responsivity* which is the ratio of its output current to the optical input power, expressed in amperes per watt.

Several devices are suitable as light detectors, these are generally *photodiodes*. They exist in several forms, e.g. the p-n junction photodiode, avalanche photodiode and the p-i-n photodiode. The latter type contains a layer of intrinsic (belonging naturally) material instead of relying on the presence of impurities for its semiconducting properties. It is also the type at present in extensive use for fibre systems and therefore the one we use as an example. It is coupled to the end of the fibre, has a response time of some $10^{-9}$ seconds and can handle modulation bandwidths of several hundred megahertz.

The intrinsic layer of a p-i-n photodiode, as the name suggests, is sandwiched between thin p and n regions as shown in Figure 6.7. Because the intrinsic (i) layer normally has no free charges, its resistance is high. Depletion layers are formed at the junctions of both p and n regions with the i-region. Hence the effective depletion layer width is increased by the addition of the i-region so that the depletion layer capacitance is reduced compared with that of a normal diode. The response of the p-i-n diode to modulated light is therefore much faster. Incoming photons entering the intrinsic layer will be absorbed and if they have sufficient energy to raise electrons across the energy band-gap between the valence and conduction levels they will create electron-hole pairs. Under reverse bias the electrons and holes separate as shown in the figure and create a photocurrent flowing in the load resistor, $R$. This is the output of the device.

P-I-N diodes have a cut-off wavelength given by:

$$\lambda = 1.24/E_g$$

where $\lambda$ is the wavelength in $\mu$m and $E_g$ is the band-gap energy in electron-volts.

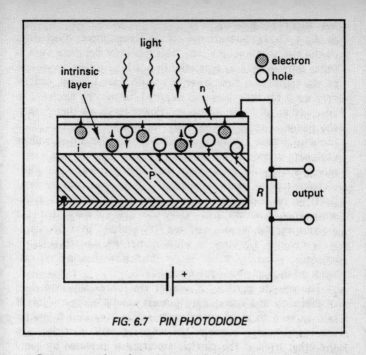

*FIG. 6.7   PIN PHOTODIODE*

Longer wavelengths cannot be detected because they have insufficient photon energy.

For silicon with a band-gap energy of 1.1 eV, $\lambda = 1.13\ \mu$m and for germanium with a band-gap energy of 0.67 eV, $\lambda = 1.85\ \mu$m. Generally silicon is used between 0.3 and 1.1 $\mu$m, germanium is suitable at longer wavelengths, e.g. 0.5 to 1.8 $\mu$m. Indium gallium arsenide is also used between about 1.0 and 1.7 $\mu$m.

As a practical example of p-i-n photodiode operation, a dark current of 30 nA may rise to as much as 60 $\mu$A at 0.95 nm with illumination no greater than we see on an overcast day.

Again we may need a reminder that whereas we have been discussing information flow in the shape of waveforms or pulses of light, each of these comprises millions of cycles of the electromagnetic wave.

## 6.4 Microwave Heating

So far we have concentrated on the communication activities of the electromagnetic wave. This must not belittle its use in other aspects of life, especially in heating. Unless very close to the transmitter the power in a radio wave is small, usually very small for little power is required for moving electrons in an antenna to create a voltage. Power levels are therefore in the picowatt range. Conversely for heating we must talk in terms of watts, kilowatts and above. From this it is evident that surrounded as we are always by a host of radio transmissions, they do us no harm with their picowatts but when electromagnetic waves are used for heating purposes we must never get in their way. This is seen in the domestic microwave cooker which has safety devices ensuring that the electromagnetic waves generated stay within. Because microwave cookers are busy in most homes we take these as an example, generally other larger microwave heating systems work on the same basic principles.

The minute particles in matter are continually in a state of vibration and accordingly possess kinetic energy. Heat is said to be a form of energy and if heat is applied to matter the added energy increases the vibration of its particles. On the other hand if the particle vibration is increased by some external means then heat is generated.

When an electromagnetic wave field encounters a dielectric the charges within the atoms and molecules are displaced and naturally the displacement follows the field polarity changes. Negative electrons are pulled one way with the positive nucleus in the opposite direction as illustrated in Figure 1.6 for one particular direction of the electric field. In this condition an atom exhibits overall polarity as shown in the figure and is called a *dipole*. Energy is absorbed from the field and heat is generated. The process is called *dielectric polarization*. High frequency reversals of the electric field result in a continual readjustment of the dipoles, the energy absorbed from the wave resulting in the generation of heat. This suggests that the faster the dipoles swing from one polarity to the opposite, the greater the amount of heat which will be generated and this is shown to be true by the simplified formula:

104

$$\text{power generated} = 2\pi f E^2 \epsilon_0 \delta \text{ watts per unit volume,}$$

showing that the power and hence heat generated is proportional to $f$, the wave frequency and also to $E^2$, the square of the electric field voltage — a result we might have expected. We are dealing with dielectrics hence $\epsilon_0$ (the electric constant — Sect.1.1.2) also appears in the formula. The symbol $\delta$ stands for the *effective loss factor* so for dielectric heating $\delta$ should be high. Fortunately water has a high value (about 16 at 3 GHz) compared with most dry materials which are often less than 0.5. Generally a frequency of 2.45 GHz is chosen for domestic ovens.

A typical microwave domestic heating system is illustrated in Figure 6.8. The electromagnetic wave is generated by a magnetron (Sect.4.2 and Fig.4.2) and is transmitted to the

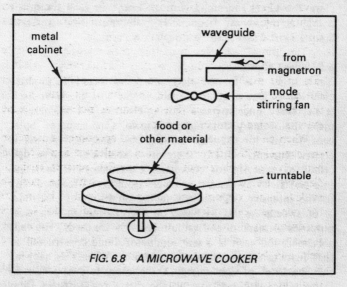

FIG. 6.8    A MICROWAVE COOKER

cabinet via a waveguide. It is projected into the closed cabinet containing the food to be heated, usually on a rotating table as shown. The food is in fact a dielectric but most of the heat produced by the wave is generated in the moisture content for it has the considerably higher value of $\delta$ as shown above.

The cabinet has internal walls of a conducting material such as aluminium or stainless steel both of which have almost zero loss factors ($\delta$) hence as shown by the formula above they absorb very little power. Most of the energy reaching the cabinet walls is continually reflected until it eventually reaches the food and is absorbed.

Also shown in Figure 6.8 is a mode stirring fan. This embodies slowly rotating blades to further project the incoming microwave energy in different directions. Doors and locks are designed so that power is disconnected from the magnetron when the door is opened and although the cooker may have a glass front, this has a metal gauze built in so that radiation from the front is prevented.

Electromagnetic wave heating (as opposed to cooking) is used extensively in the medical world. It is called "diathermy" from the Greek, meaning *through heat*. In such treatment energy at microwave frequencies is transmitted into the skin from a short distance to heat deep down muscle.

## 6.5 Solar Cells

Yet another role for the electromagnetic wave is to produce electricity. One might reasonably argue that the wave itself is electricity since it has a moving electric field but here we mean the ordinary stuff measured in volts and amperes. Solar cells do this for us, their input is light and their output is electric current. Briefly therefore a solar cell can be described as a device for converting electromagnetic radiation into electricity by generating an electromotive force according to the intensity of the light falling on its input electrode.

Solar cells are usually based on silicon although for special purposes germanium and gallium arsenide are used. The cells are small and each is a semiconductor diode constructed so that light can impinge on the junction. When this happens the "packets" of light energy, the photons (see Appendix 5) in colliding with electrons in the diode release them from their atom orbits and create electron-hole pairs, i.e. free electrons each paired with one atom which is minus one electron in its outer orbit. Note however that electrons can have only certain energy levels and on colliding with a photon the latter must provide sufficient energy to the electron for it

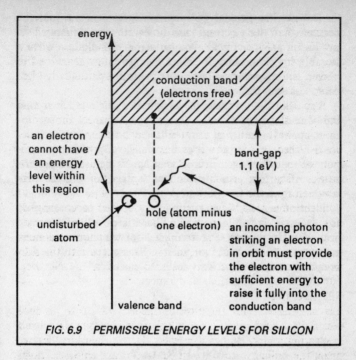

*FIG. 6.9   PERMISSIBLE ENERGY LEVELS FOR SILICON*

to be raised energy-wise into the conduction band, i.e. to be released from its parent atom. If this condition does not apply the radiation is not absorbed.

Taking silicon as an example, the minimum extra energy which a photon must provide when it collides with an electron in orbit so that the electron can be released is 1.1 eV as shown in Figure 6.9. From Planck's basic relationship (Sect.6.3.2) this corresponds to a wavelength, $\lambda = 1.24/E_g$ $\mu$m ($E_g$ in electron-volts).

Hence for $E_g = 1.1$ eV, $\lambda = 1.13$ $\mu$m and from Figure 6.5 this is in the infra-red. Clearly then light at wavelengths longer than this will not contain photons possessing sufficient energy for the release of electrons from their parent atoms in silicon. This is found to be approximately so in practice and generally silicon solar cells operate from this wavelength down to around 0.4 $\mu$m.

The electric field of the depltion layer at a p-n junction ensures that the electrons and holes are kept separated so producing an open-circuit voltage across the diode. When a suitable load resistance is connected current can flow and so power can be extracted. Efficiencies are not particularly high, often less than 20%.

Typically the open-circuit voltage of a single cell is around 0.5 V and the short-circuit current 0.1 A, hardly enough for most applications hence many cells may be connected in series for a higher voltage output or in parallel for a higher current. Series-parallel arrangements on *solar panels* are used to provide sufficient current at a high enough voltage for general semiconductor circuit use, e.g. 12 V at 50 mA with fairly bright sunlight at say, 1 kW per square metre. Larger panels are used for example in satellites where there is plenty of sunshine around or in remote terrestrial areas for telephone and radio equipment. The banks of solar cells are generally used in conjunction with secondary cells to maintain the electrical supply throughout periods of darkness.

# AND FINALLY

The author sincerely hopes that you, the reader have found these few notes interesting and absorbing. The text may have raised more questions than it has answered but this must be the way of things confronted as we are with an invisible subject of such extreme complexity.

At least we have got to grips with some of the vagaries of the electromagnetic wave over most of the radio spectrum and we may even have seen the light about light.

# Appendix 1

## CONDITIONS FOR RETURN OF AN ELECTROMAGNETIC WAVE FROM THE IONOSPHERE

It has been estimated that within the F region, provided that the collision frequency is low, the refractive index, $n$ can be expressed approximately by:

$$n = \sqrt{1 - (Ne^2/\epsilon_0 m_e \omega^2)}$$

where $N$ is the number of electrons per m³
$e$ is the electron charge ($1.6022 \times 10^{-19}$ coulombs)
$\epsilon_0$ is the electric constant ($8.8542 \times 10^{-12}$ Farads/metre)
$m_e$ is the electron mass ($9.1095 \times 10^{-31}$ kg)
$\omega = 2\pi \times$ frequency of transmission ($f$).

Substituting the practical values of $e$, $\epsilon_0$ and $m_e$ leads to an approximate relationship known as the Appleton formula (after Sir Edward Appleton):

refractive index of the ionosphere:

$$n = \sqrt{1 - (80.6 \, N/f^2)}$$

At the bottom of the ionosphere the electron density is zero hence from Snell's Law ($n_1 \sin \theta_1 = n_2 \sin \theta_r$ — Sect.3.2). $n_1 = 1$ and considering the condition where the wave is caused to move horizontally:

$$\theta_r = 90° \qquad \therefore \sin \theta_r = 1$$

hence the refractive index, $n_2 = \sin \theta_1$.

Accordingly the wave reaches its highest point in the ionosphere when:

$$\sqrt{1 - (80.6 \, N/f^2)} = \sin \theta_1$$

or equally, since $\sin^2 \theta_i + \cos^2 \theta_i = 1$:

$$N = \frac{f^2 \cos^2 \theta_i}{80.6}.$$

This value of $N$ turns a wave so that it travels horizontally. If the electron density is greater then the wave will be bent over sufficiently for a return to earth. If the electron density is less, the wave will not be returned at this point but will continue to travel upwards to the level at which the value of $N$ satisfies the equation above. If such a value of $N$ is not experienced, the wave is not returned.

To reflect a wave arriving at normal incidence requires the greatest electron concentration. To calculate the highest frequency which when arriving at normal incidence will be returned ($\theta_i = 0$), then from the above equation, since $\cos^2 \theta_i = 1$:

$$f_c = \sqrt{80.6\,N}$$

where $f_c$ is known as the *critical frequency*.

As an example, a 10 MHz wave reaching the $F_1$ layer in the ionosphere at an angle of $70°$ to the normal requires an electron concentration for the wave to be returned to earth of:

$$N = \frac{(f^2 \cos^2 \theta_i)}{80.6} = \frac{10^{14} \times 0.117}{80.6}$$

$$= 1.45 \times 10^{11} \text{ electrons per m}^3,$$

a concentration well within the range for the $F_1$ layer as seen from the estimates below. These give an idea of the range of electron concentrations in the ionosphere — but never forget the variability!

$F_2$ layer at 500 km:

$10^{11} \rightarrow 5 \times 10^{11}$ electrons per m$^3$

$F_1$ layer at 200 km:

$2 \times 10^{11} \rightarrow 4 \times 10^{11}$ electrons per m$^3$

E layer at 120 km:

$10^{11} \rightarrow 2 \times 10^{11}$ electrons per m$^3$

D layer at 80 km:

average around $10^9$ electrons per m$^3$.

# Appendix 2

# PROPAGATION OF RADIO WAVES IN FREE SPACE

It may come as a surprise to some readers to find that free space, just nothing at all, has an impedance. Mathematically this can be shown to be so and we find that it has no quadrature component (i.e. is non-reactive), hence no phase change is imposed on a wave travelling through. The dimension of the electric field of a wave ($E$) is in volts per metre and that of the magnetic field ($H$) (which is associated with current flow) is in amperes per metre, hence by Ohm's Law their ratio has the dimension of impedance. Accordingly the impedance of free space as seen by a wave travelling through it is:

$$Z_0 = E/H \text{ ohms} .$$

It can also be shown that:

$$Z_0 = \sqrt{\frac{\mu_0}{\epsilon_0}}$$

where $\mu_0$ is the permeability of free space (Sect.1.1.4) and $\epsilon_0$ is the permittivity of free space (Sect.1.1.2), i.e.

$$Z_0 = \sqrt{\frac{4\pi \times 10^{-7}}{8.854 \times 10^{-12}}} = 377 \ \Omega .$$

(If the value for $\epsilon_0$ of $10^{-9}/36\pi$ is used instead, this results in a sometimes more convenient figure for $Z_0$ of $120\pi$ ohms.)

$Z_0$ is also known as the *intrinsic* (belonging naturally) impedance of free space.

We also know that electromagnetic waves travel enormous distances in free space and do not appear to suffer attenuation at all. This can be demonstrated mathematically. The propagation loss through *any* medium is indicated by its attenuation constant, $\alpha$. This factor is normally expressed in

113

decibels per metre and the basic equation is:

$$\alpha \simeq \frac{\sigma}{2} \sqrt{\frac{\mu}{\epsilon}}$$

where $\sigma$ is the *electric conductivity* of the medium and clearly for free space, $\sigma = 0$, hence $\alpha = 0$ showing that an electromagnetic wave is not attenuated in its passage through free space. When "space" is not completely free, i.e. even a few molecules of gas are around as in the ionosphere, the wave sets these in motion. To do this energy is extracted from the wave, hence it is attenuated.

# Appendix 3

## THE ISOTROPIC ANTENNA

This is a purely theoretical transmitting antenna producing the same radiation in all directions from a point source. "Isotropic" is derived from Greek meaning *equal in all directions*. The concept is useful for assessment of the gains of practical antennas. Figure A3.1 shows the idea pictorially,

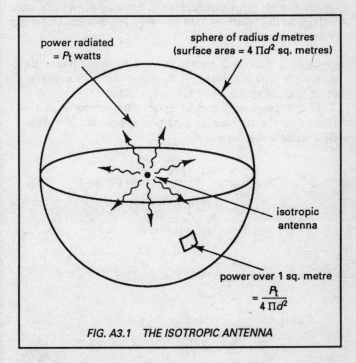

power radiated
= $P_t$ watts

sphere of radius $d$ metres
(surface area = $4 \Pi d^2$ sq. metres)

isotropic
antenna

power over 1 sq. metre

$= \dfrac{P_t}{4 \Pi d^2}$

**FIG. A3.1   THE ISOTROPIC ANTENNA**

concluding that for such an antenna radiating a power of $P_t$ watts, the power flux density (p.f.d.) at a distance $d$ metres:

p.f.d. $= P_r = P_t/(4\pi d^2)$ watts per square metre (W/m$^2$).

Also:

$$P_r = E^2/Z_0 \quad \text{(Sect.2.3)}$$

where $E$ is the wave electric field strength and $Z_0$ is the intrinsic impedance of free space ($120\pi$ ohms). Then:

$$E^2 = P_r \times Z_0 = \frac{P_t}{4\pi d^2} \times 120\pi = \frac{30P_t}{d^2}$$

$$\therefore E = \frac{\sqrt{(30P_t)}}{d} \quad \text{volts per metre.,}$$

hence at a distance $d$ metres from an isotropic transmitting antenna radiating a power of $P_t$ watts, the field strength, $E$ of an electromagnetic wave can be calculated.

For a practical transmitting antenna having a gain $G_t$ relative to the isotropic in a particular direction, the electric field strength is given by:

$$E = \frac{\sqrt{30G_t \times P_t}}{d} \quad \text{volts per metre.}$$

# Appendix 4

## THE ELECTRON-VOLT

When we are considering the energy of a single or a few electrons, the SI unit of energy, the joule is sometimes found to be inconveniently large, accordingly one referring to a single electron is frequently used.

When an electric field accelerates a single free electron through a potential difference of one volt, the energy acquired by the electron due to its higher velocity is said to be one electron-volt (symbol eV). Hence work done by the electric field = $e \times V$ where $e$ is the electron charge ($1.602 \times 10^{-19}$ coulombs) and $V$ is the voltage through which the electron has been accelerated. Therefore:

$$1 \text{ eV} = 1.602 \times 10^{-19} \text{ coulomb-volts (joules)}.$$

Note that although "volt" appears in the name of the unit, the electron-volt is a unit of energy, not voltage.

# Appendix 5

## THE PHOTON

Most of us are quite satisfied with the electromagnetic wave theory of light as an explanation of its activities until we meet photoelectricity. For this something else is needed and it was Einstein who originally used Planck's ideas on radiation to explain the photoelectric effect. This leaves us with seeing light both as having a wave nature but also as having a particle nature in that it behaves as though it consists of infinitesimally small particles which we call *photons*. However it never seems to behave as both at the same time. Hence we have to see light according to which theory suits it better. More technically the two theories are known as the electromagnetic theory and the quantum theory and at times we may have to consider light either as having a wave or a particle nature.

The quantum theory suggests that light propagates in separate *quanta* or photons and a single photon energy $(E)$ is related to the light frequency $(f)$ by:

$$E = hf \text{ joules}$$

where $h$ is Planck's constant equal to $6.626 \times 10^{-34}$ joule-seconds.

To get a feeling for the size of a photon, consider a fibre optic system operating at 0.6 $\mu$m (orange light — Fig.6.5) with an incoming optic power of 1 $\mu$W. We need to know the number of photons arriving in 1 $\mu$s.

Firstly convert the transmission wavelength into its frequency:

$$f = c/\lambda = 3 \times 10^8/0.6 \times 10^{-6} = 5 \times 10^{14} \text{ Hz}$$

then:

$$E = hf = (6.626 \times 10^{-24}) \times (5 \times 10^{14}) = 3.3 \times 10^{-19}$$

joules. This is the energy possessed by a single photon in a

0.6 $\mu$m wavelength transmission.

Total energy arriving in $10^{-6}$ seconds = 1 $\mu$W × $10^{-6}$ secs = $10^{-12}$ joules. Therefore the number of photons per *microsecond* is equal to $10^{-12}/3.3 \times 10^{-19} = 3.03 \times 10^6$.

Sensitive photodetectors can easily recognize this number of photons arriving, in fact it is possible to detect radiation levels in which only a small number of photons is involved.

# Index